SERVICE INDUSTRIES IN THE WORLD ECONOMY

P. W. Daniels

BLACKWELL
Oxford UK & Cambridge USA

© P. W. Daniels 1993

The right of P. W. Daniels to be identified as author of this work has been asserted in accordance with the Copyright, Designs and Patents Act 1988.

First published 1993

Blackwell Publishers
108 Cowley Road
Oxford OX4 1JF
UK

238 Main Street
Cambridge, Massachusetts 02142
USA

British Library Cataloguing in Publication Data

A CIP catalogue record for this book is available from the British Library.

Library of Congress Cataloging-in-Publication Data

Daniels, P. W.
　　Service industries in the world economy / P. W. Daniels.
　　p.　　cm.
　　Includes bibliographical references and index.
　　ISBN 0–631–17703–5. — ISBN 0–631–18132–6 (pbk.)
　　1. Service industries.　I. Title.
HD9980.5.D363 1993　　　　　　　　　　　　92–43005
338.4--dc20　　　　　　　　　　　　　　　　　CIP

This book is printed on acid-free paper.

Typeset in 11 on 13 Plantin
by Photoprint, Torquay, Devon.
Printed in Great Britain by Page Bros. Ltd., Norwich

Service Industries in the World Economy

WITHDRAWN

IBG STUDIES IN GEOGRAPHY

General Editors
Felix Driver and Neil Roberts

IBG Studies in Geography are a range of stimulating texts which critically summarize the latest developments across the entire field of geography. Intended for students around the world, the series is published by Blackwell Publishers on behalf of the Institute of British Geographers.

Published

Debt and Development
Stuart Corbridge

The Changing Geography of China
Frank Leeming

Critical Issues in Tourism
Gareth Shaw and Allan M. Williams

The European Community
Allan M. Williams

In preparation

Geography and Gender
Liz Bondi

Population Geography
A. G. Champion, A. Findlay and E. Graham

Rural Geography
Paul Cloke

The Geography of Crime and Policing
Nick Fyfe

Fluvial Geomorphology
Keith Richards

Russia in the Modern World
Denis Shaw

A Geography of Housing
Susan Smith

The Sources and Uses of Energy
John Soussan

Retail Restructuring
Neil Wrigley

Contents

Tables

Figures

Preface

The geographical domain of service industries now extends far beyond local or national boundaries to embrace the international stage. This book attempts to identify the significant processes that have enabled some service industries to internationalise at rapid rates in recent years. The situation is complex because service industries such as telecommunications have enabled the process of internationalization to take place, while themselves being active participants; others such as business and professional services could not have become involved without the support role performed by information technology services. During the 1980s service industries have become a much more active ingredient in shaping economic and social change at the global scale; they have also stimulated national governments and other organizations and agencies involved with the regulation and promotion of international trade to look more carefully and constructively at ways in which the benefits can be as evenly distributed as possible between the developed and less developed economies. The resulting changes have stimulated foreign direct investment in services and a demand for new and diversified service products that reflect the growing flexibility of cross-border transactions, labour flows, investment opportunities and sophisticated innovations in ways of making finance capital available for international corporate growth or production strategies.

But there are spatial variations in factor endowments between the world's nations, regions and cities that ensure that their ability to supply services, as well as to generate demand, is highly uneven. These differences are accentuated by wide variations in the attitudes towards services by national governments as reflected in the rules and regulations governing the operation of foreign enterprises.

Some nations are more open than others; some are suspicious of the nations that are advocating greater liberalization of trade in services, others are very anxious to protect their domestic service industries since, until relatively recently, they were confined to national markets. There are major contrasts, for example, in the extent to which the developed, less developed and post-socialist economies of Eastern Europe already participate in the global service economy, as well as the opportunities open to them in the future. The most potent symbols of the uneven pattern of international development of services are the world's major cities; most have shared in the general process of tertiarization of their economic infrastructure but a relatively small number have attracted a disproportionately large share of the benefits (as well as the costs).

The book is organized into six chapters. Chapters 1 and 2 outline some of the reasons for expansion of services into the international arena. Particular emphasis is given to the circumstances that have contributed to the tradability of services, notably the central role performed by technological change, the emergence of service multinational enterprises and the part played by changes to the regulatory environment for the operation of service industries. It is of course extremely important to try to capture the extent to which service industries are an integral part of the world economic system. Inevitably, it is necessary yet again to bemoan both the poor quality and the limited depth of national and international service industry statistics. That statistics are out of date, that they are insufficiently disaggregated by sector or by geographical units, and that it is difficult to compare different series because of definitional inconsistencies, are just some of the standard laments that continue to undermine serious efforts to measure and understand the international role of service industries. International organizations such as the United Nations Conference on Trade and Development (UNCTAD), the International Monetary Fund (IMF), the General Agreement on Tariffs and Trade (GATT), the United Nations Centre on Transnational Corporations (UNCTC) and the Organization for Economic Co-operation and Development (OECD) have gathered some statistics. These are used in chapter 3 to examine the extent of global trade and the characteristics of foreign direct investment by services. This distinction is crucial since it helps us to understand how the process of globalization in services is different to that in manufacturing.

Chapters 4 and 5 develop the argument that the uneven pattern of international trade in services, together with the concentrated distribution of foreign direct investment, is exerting considerable influence on the form and development of the global urban system. A small number of networked (or world) cities at the top of the urban hierarchy are the key players in the global service economy; they act as control points for the flows of information, knowledge and investment that shape the economic and social prospects of distant cities, regions and even nations. The geographical concentration that has accompanied the internationalization of services has also contributed to the internal restructuring of some of the world's leading cities, and some example are outlined briefly. The final chapter offers some reflections on the future role of services in the world economy: can the trends evident during the 1980s be assumed to shape the evolving geography of services during the 1990s and beyond?

Although we live in an environment in which information is ever more readily accessible via the wonders of modern technology, writing a book seems to have changed little. It remains a slow and tortuous process punctuated by encouragement from friends and family and a heavy dependence on inspiration from the research and writing of others. I am particularly indebted to Nigel Thrift for initially suggesting that a book on this subject would be useful; he has subsequently shown great patience as the deadline for the manuscript inexorably slipped past. The possibilities emerged during our joint research during the mid 1980s (Andrew Leyshon, now at the University of Hull, also played a major part) on the growth and location of professional producer services in the late 1980s. We were surprised by the range and diversity of the internationalization activity that we uncovered; there was clearly scope for a more general analysis of the global development of service industries. A number of colleagues who heard a paper based on part of this book at a Fulbright Commission Colloquium on 'Buying back America: UK capital in the UK service economy' in Cardiff in December 1990 made some useful comments on what needed to be done and identified some omissions, especially relating to statistical sources. I am also grateful to two anonymous readers of the final draft of the manuscript for their constructive comments and suggestions on the structure and content of the book. I have tried to rework parts of the manuscript to take their comments into account. Ultimately, of course, I take full responsibility for the final product.

Carol Smith and Shirley Passey in the Department of Geography, University of Portsmouth coped patiently with the typing and retyping of different parts of the manuscript; they both provided an excellent service! My distinct lack of computer literacy led to considerable dependence on advice from Mark Bobe of the Service Industries Research Centre at the University of Portsmouth, who also very kindly provided invaluable assistance with library searches for reference and statistical material as well as with an ESRC funded project which has provided some of the material included in chapter 5.

I do not know why most authors leave their family to the end of their list of acknowledgements: Carole, Paul and Charlotte have been, as on previous occasions, a constant source of that blend of love, humour and understanding that keeps the demands of the omnipresent word processor in perspective. Trying to convince a sceptical son reading classics at the University of Durham how geography, service industries and the world economy are a legitimate part of a discipline which 'is all about capes and bays' (his perception) has been an amusing diversion over the last two years!

P. W. Daniels

ONE

The Rise of Services: Some Factual and Theoretical Perspectives

Introduction

Service industries have long been the Cinderella of economic geography. It has been suggested that the 'service sector is one of the least understood portions of our global economy' but that 'no economy can survive without a service sector' (Riddle, 1986, 1, 2). A recent report for the General Agreement on Tariffs and Trade (GATT, 1989) notes that services such as insurance, banking, telecommunications and transport reach to the heart of national economies and provide essential inputs to manufacturing. Put another way, there are 'forward and backward linkages that make the service sector so central to economic planning. Services are the glue that holds any economy together, the industries that facilitate all economic transactions, and the driving force that stimulates the production of goods' (Riddle, 1986, 26). Dicken (1992, 349) has described the recent internationalization of services as a process that has contributed to 'making the world go round'. As we shall see from the evidence outlined in this volume, this is a neat analogy which Dicken elaborates with reference to the strategies and development of financial services in global markets during the 1980s but which can also be illustrated in relation to the behaviour of other types of services.

Countries which want to compete in world markets need to foster or to acquire these services as cheaply, quickly and efficiently as possible. Service industries have therefore been receiving rather closer attention from academics, planners, policy makers and even politicians. Service industries now account for at least two-thirds of total employment in developed economies and for at least 50 per cent of gross domestic product (GDP). They make a less significant contribution at present to economies of the developing countries or of Eastern Europe. But as the latter experience the transition from a centrally planned to a market economy it is certain that services will have a prominent role. At another level national economies are becoming less insular and more interdependent in relation to services, a process that is inextricably linked to the effects of the information technology revolution and to the ways in which firms are organized and business is transacted. Such 'globalization' refers to 'the interlinkage that is established among national markets as a result of the progressive removal of exchange controls on capital flows and the liberalization of various financial centres in developed countries together with the revolution in information gathering, processing and transmission made possible by the technological developments in the fields of computing and telecommunications' (Bhatt, 1989, 153; see also Knox and Agnew, 1989). Thus, Harvey (1982, 373) observed that: 'Monetary relations have penetrated into every nook and cranny of the world and into almost every aspect of social, even private life . . . and this . . . has also been accompanied by physical transformations that are breathtaking in scope and radical in their implications.' One of the principal indicators of the transformation that is taking place is the marked increase in the production, consumption and trade in services (Ochel and Wegner, 1987).

Such changes are not without geographical consequences: they are influencing the relative growth of nations, regions and cities around the globe. Places that were prominent participants in the growth that accompanied the Industrial Revolution have not, and will not necessarily, maintain their status and economic power in what is often termed the post-industrial (Bell, 1973; Hirschorn, 1988) or information society. While the economic wealth that was produced by these places was almost certainly dependent to some degree on international trade in raw materials and merchandise goods, their current fortunes are increasingly bound up with international flows of people and services as well as goods. The more traditional, low-key role of services in the development process is undergoing a transformation. Services that are based on the production and delivery of knowledge and information are acknowledged as having a key role in the ability of countries to adapt and apply advanced technologies to the production process, to innovations in production and to improved efficiency, and ultimately to be able to respond more effectively to competitive forces in global markets. Variously called information or producer services, they contribute to, and help to enhance, the human capital of countries; indeed trade in these services is becoming as important for attracting foreign exchange as trade based on merchandise goods.

This book attempts an exploration of some of the global dimensions of the development of service industries, as well as of questions such as: how have service industries grown and diversified in developed, developing and centrally planned economies? What are the factors that help to explain any observed differences? What are the patterns of trade in services and how can we understand these patterns? Where is the expansion (or indeed the contraction) of economic activity linked with services located? How can the interdependence of national economies or places in relation to service industries be illustrated? How have different places handled the demand for new infrastructure, buildings and occupational skills, for example, that has been created by the expansion of the service sector.

Services: Definition and Classification

Before proceeding, we should dwell briefly on the question: what are services, and how are they classified? This is very important, and there are almost as many answers as there are researchers that have written on the subject (see for example Marshall et al., 1988; Singelmann, 1978; Fuchs, 1968; Stanback, 1979). A useful way to define a service is on the basis of the utility that it provides: some yield immediate or short-term utility (fast-food restaurant, cinema, launderette, petrol station); others offer medium-term or semi-durable utility (automobile repair, taxation advice, dental treatment); yet others are more durable and provide much longer-term utility (mortgage financing, life insurance, pension arrangements). All of these are levels of utility that are particularly applicable to final consumption. But a similar notion could be applied to intermediate consumption: short-term utility is provided by office cleaning services, express document delivery, foreign currency dealing or a share price information service; medium-term utility is provided by an office market report prepared by a property consultant or a market profile for an individual country by an economic consultant; longer-term utility is provided by advice on the installation of a computer network, the design and construction of a commercial building or the management of a resource such as water supply. These examples illustrate the elusiveness of services: they are not tangible in the way typical of the goods produced by a manufacturer. Yet an automobile, although obviously very tangible, also provides a service to its owner. The 'value' of this service will depend on the way in which the automobile is used, the ability of the owner to maintain it in running order, and the resources available to use it as and when required. Therefore, service activities 'connect agriculture and manufacturing with the consumption of goods and services, and with the management of organizations and institutions of society' (Castells, 1989, 129).

It will be apparent that services are diverse and consequently difficult to classify. A grouping of services into those meeting final demand and those meeting intermediate demand is widely used (Marshall et al., 1988). The so-called consumer and producer services are not, however, mutually exclusive groups; some service activities such as banking, insurance and finance fulfil both final and intermediate demand. One solution is to create a third group of 'mixed' services or to assign an activity to the consumer or producer group according to which kind of output (using input–output tables) predominates. A related possibility is to estimate the number of employees contributing to final or intermediate output and assign the activity accordingly.

Although these methods have been widely adopted, there remains a good deal of disquiet about their value as the framework for studying service industries (Walker, 1985; Allen, 1988). It has been suggested (Allen, 1988) that this 'technical' solution is too neat and is conceptually weak. It fails to take account of the way in which the balance between final and intermediate output for any one service will fluctuate over time and space. In addition, it is misleading to classify banks, for example, as a producer service; the value of deposit and money transactions far outweighs the value of loans at a bank, but the majority of the employees are engaged in work connected with the latter. This leads Allen (1988, 18) to propose that some services are neither consumer nor producer services but are circulation services:

> Whereas advertising, marketing and research and development are readily identifiable as producer services and, similarly, recreational, education, health, welfare and personal services fall neatly under the category of consumer services, the commercial and financial services which mediate and abbreviate the exchange process within an economy are neither producer nor consumer services. They are circulation services, services produced *within* the process of circulation and *for* circulation, and not intermediate services produced primarily for other branches of industry or final services produced for consumers.

This is a useful way of reducing some of the limitations of the consumer/producer dichotomy and certainly helps with the task of analysing the processes responsible for the globalization of service industries (see chapters 2 and 3).

Another classification of services that is now widely accepted was originally proposed by Singelmann (1978). Four groups are identified: distributive services (e.g. retail, transportation); producer services (e.g. insurance, banking, engineering, legal services); social services (e.g. education, medical, welfare, government); and personal services (e.g. hotels, laundry, entertainment). This is a valiant attempt to impose some order on the diversity of the service industries while also clearly separating them from the transformative (manufacturing and construction) industries and the extractive activities (agriculture and mining). But this is also problematical because it presupposes that the service sector is somehow exclusive; in practice the boundaries between it and the other sectors are becoming blurred as some manufacturers also produce services (see chapter 2). Taken to its extreme we might conclude that 'there is not a service sector' (Castells, 1989, 130); rather there is a set of activities that have increased in diversity or specialization as society has evolved, and services (especially social and personal) are 'in fact, a way of absorbing the surplus population generated by increased productivity in agriculture and industry' (Castells, 1989, 130). The implication is that if productivity deteriorated in manufacturing, the service sector would contract. The indivisibility of services and manufacturing in this way has been advocated by Cohen and Zysman (1987) amongst others.

The definitional and classification problems that have plagued the study of services will persist. Technology, for example, has been responsible for generating new service occupations that are often linked with new products provided by enterprises which could not have existed even 25 years ago. Examples are computer and software services or specialized information services that are totally reliant on telecommunications. These have been amongst the activities that have contributed to the globalization of services that is the principal theme of this book. Their activities are not totally removed from the behaviour of manufacturing but they have recently been engaged in a process which is notable for its pace of development and its implications for the global geography of services. It therefore seems worthwhile to consider services as a separate entity for the purpose of this book while acknowledging some of the difficulties that this approach entails. Hopefully, this will not undermine an apprecia- tion of the processes under way and which are largely service led (Riddle, 1986).

The Recent Expansion of Services

It is essential to begin with an outline of the principal attributes of the recent growth of service industries. Ideally, this evidence should be presented for a selection of countries across the spectrum from the developed, to the less developed and the centrally-planned economies. In practice, the statistics that are available make this approach somewhat difficult; there are enough problems in obtaining data for comparable countries before the added com- plexity of attempting to standardize comparisons across groups of countries is encountered. Most published studies therefore confine their analyses to one group of countries only and rarely attempt direct comparison between groups.

Employment growth in services is most readily charted for developed countries (see for example Elfring, 1988; Ochel and

Table 1.1 Service employment[a] as a percentage of total employment, developed market economies, 1960–1985

Year	France	Germany	Japan	Netherlands	Sweden	United Kingdom	United States	Average
1960	44.1	38.6	37.4	47.8	47.7	48.8	61.1	46.5
1973	51.3	46.1	49.1	57.7	57.7	55.4	66.4	54.8
1985	61.7	54.2	57.1	67.5	66.9	65.3	72.3	63.6

[a] Service employment defined by the International Standard Industrial Classification: wholesale and retail trade and restaurants and hotels (division 6); transport, storage and communication (division 7); financing, insurance, real estate and business services (division 8); community, social and personal services (division 9).
Source: extracted from Elfring, 1989b, 415, table 2

Wegner, 1987; Noyelle and Stanback, 1988; Kellerman, 1985; Cuadrado and Del Rio, 1989; Illeris, 1989). Without exception, in the seven developed market economies shown in table 1.1, service employment has increased its share of total employment between 1960 and 1985. The United States heads the list as the country with a consistently above-average share of service industries in all employment – more than seven out of every ten jobs in 1985. At the other end of the spectrum, Germany has consistently remained below average with only just over one in two jobs in services in 1985. Japan also has a lower proportion of its employment in services than might be expected from its prominent role in the modern global economy. The relatively low proportion of services in these two countries may be connected with the major role of the construction industries during reconstruction and readjustment following World War II (a phenomenon that may also be important in Italy: (UNCTAD, 1989a). Construction employment had already begun declining in the United States by 1970.

The increasing share of services in developed economies coincided between 1960 and 1973 with improvements in individual and national prosperity, relatively low unemployment rates and sharp increases in output and productivity in the manufacturing industries (UNCTAD, 1989a). Unemployment rates increased and productivity improvements slowed down between 1973 and 1985, yet the services share of total employment continued to increase. The ability of services to continue expanding during periods of economic recession or slower economic growth is often explained by

Table 1.2 Changes in the share of service employment in total employment, by subsector, developed market economies, 1960–1973 and 1973–1985

Subsector[a] and period	France		Germany		Japan		Netherlands		Sweden		United Kingdom		United States		Average	
Producer																
1960–73	2.5[b]		1.8		3.2		2.6		1.6		2.1		2.3		2.3	
1973–85	2.5[c]	8.5[d]	2.0	7.2	3.1	9.6	3.8	10.5	1.3	0.4	3.0	9.5	3.9	12.6	2.8	9.2
Distributive																
1960–73	1.8		0.6		4.8		0.1		0.4		−0.5		−0.7		1.0	
1973–85	1.4	20.0	−0.1	18.0	1.5	24.8	0.6	21.1	−0.7	19.1	1.2	21.3	−0.1	21.4	0.5	20.8
Personal																
1960–73	−0.4		−1.1		1.2		−0.9		−1.8		−0.1		−0.4		−0.4	
1973–85	0.2	7.7	1.3	7.8	1.0	9.9	0.9	8.3	−0.5	6.1	2.0	9.9	1.5	12.4	0.9	8.9
Social																
1960–73	3.2		4.0		2.3		8.1		9.9		5.0		4.1		5.6	
1973–85	6.5	25.7	5.0	21.3	2.2	12.7	4.8	27.6	9.1	35.3	3.8	24.6	0.5	25.8	4.5	24.7
Total services																
1960–73	7.2		9.5		11.7		10.1		10.0		6.6		5.3		8.3	
1973–85	10.4	61.9	8.1	54.3	8.0	57.0	9.8	67.5	9.2	60.9	9.9	65.3	5.9	72.2	8.8	63.6

[a] Producer services, ISIC division 8; distributive services, ISIC divisions 6 and 7; personal services, ISIC division 9 (part); social services, ISIC division 9 (part).
[b] Difference between shares of total employment in 1960 and 1973.
[c] Difference between shares of total employment in 1973 and 1985.
[d] Percentage of total employment in 1985.

Source: derived from data in Elfring, 1989b, 415, table 2

reference to their inferior rates of improvement in productivity (see for example Gadrey, 1988; McKinsey Global Institute, 1992). While the debate about how to measure the output and productivity of service industries continues, it would be unwise to rely solely on these indices to explain the shift that has been taking place steadily since the 1970s. Another index is the increase in the demand for services that support and improve the quality of life and for services that raise the educational standards and skills of the labour force in developed economies. Explanations can also be couched in terms of the changing characteristics of production, i.e. an increasing share of factor inputs is derived from the knowledge, information and expertise that are incorporated in the design, manufacture, packaging, marketing, advertising and distribution of goods and services. Furthermore, the administrative and organizational structures that are required to support these new production methods also generate a requirement for new services.

One way of demonstrating these effects is to examine changes in the shares of service employment by subsector 1960–73 and 1973–85 in developed market economies (table 1.2). Two features are prominent: first, the large proportion of total change in both periods that has arisen from the expansion of employment in the social services; and second, the positive contribution of producer services during both periods (even though at a lower level than for social services). The emergence of producer services has arisen, as we shall see in more detail later, from changes in the way that goods and services are produced and distributed as well as from changes in the way in which the fiscal resources used to pay for these activities are utilized, i.e. the circulation of finance capital. Producer services have become more closely integrated in production – a trend that coincided during the 1980s with a number of advanced economies having large surpluses of national income, most notably in the Netherlands and Sweden. This permitted large-scale investment in the public provision of education, health, social and community services. Some of these investments are part of deliberate economic policy initiatives designed to combat structural unemployment; others are part of a commitment to the use of subsidies to ensure equitable access to welfare services. In other developed market economies such as the United States the provision of social services

Table 1.3   Service employment[a] as a percentage of total employment, selected developing countries, 1977–1984

Year	Egypt	Kenya	Zambia	Brazil	Jamaica	Bahrain	Pakistan	Average
1977	35.3	51.3	46.4	9.0	46.7	55.9	26.4	11.2
1980	35.7	55.4	48.5	46.1	46.9	61.1	26.4	25.3
1984	36.2	59.0	49.7	47.1	49.1	61.5	27.5	43.9

[a] Service employment defined as in table 1.1.
Source: International Labour Office, 1987

is more closely linked to the market, with the result that the changes in the level of social services employment are more limited.

While the share of service employment in most of the developing countries has yet to match the levels demonstrated by the developed market economies (table 1.3) there is also more diversity. In most, however, the share of services in total employment in 1977 was still lower than the equivalent figure for developed market economies; subsequent increases in shares have also been markedly slower. By 1984 only a few 'special cases' such as Singapore, Bahrain or Colombia had levels of service employment approaching those typical of the developed countries in 1985 (see table 1.1). For some developing countries such as Egypt, Pakistan or Thailand the 'services gap' remains very large. However, in common with developed market economies, employment in social and personal services had expanded to form a significant share of total service employment in 1984: 77 per cent in Jordan and 71 per cent in Kenya according to the International Labour Office (1987) (ILO). Distributive services are normally the next largest employers, followed by transport and communications and the financial and business services sector. A notable feature of the statistics for the latter is their incongruous diversity: according to the ILO, one in four service sector jobs in Nigeria is in the financial and business services, compared with just 11 per cent and 8 per cent in Singapore and Bahrain respectively. The latter are acknowledged as second-tier international financial centres but Nigeria is not. This highlights the difficulty of gathering comparable statistics and, therefore, the need to use the available data very cautiously when comparing the scale and structure of services in developing countries or making comparisons with developed market economies.

The contribution of the service industries to GDP (45–55 per cent) also tends to be lower in developing countries, noticeably so in the use of producer services. Whereas some 20 per cent of the GDP of developed market economies is generated by producer services, the equivalent figure for low-income developing countries is nearer 5 per cent (UNCTAD, 1989a) and struggles to exceed 10 per cent in upper-income developing countries. Another important difference between developed and less developed economies is the role of the informal sector in the production and distribution of services. A wide range of services, including retail and some producer services, can function via the informal sector. Although not all the activities involved are of low skill, the net effect of production through the informal sector is to undermine their potential contribution to formal economic growth and development. As we will see later, these two attributes of service industries in developing countries make them vulnerable to international competition and tend to undermine their actual and potential contribution to the development of other indigenous economic activity.

The rapid changes in the political and economic organization of the centrally planned economies of Eastern Europe since 1989 will undoubtedly lead to greater integration with the economies of Western Europe and with countries further afield. This has increased awareness as well as anxiety about the role that services will play in this process (Maciejewicz and Monkiewicz, 1989; Schneider, 1991; Nesporova, 1990). Until very recently most of the service industries in the centrally planned economies were perceived as non-productive, in line with the tenets of the so-called Smith-Marx-Lenin theory. Accordingly, capital formation and labour specialization are the bedrock of national economic growth; manufacturing is the centrepiece, with services only performing a supporting or dependent role. The effect of this, also evident in countries more recently organized along the same lines, such as Cuba, has been an underestimation of the contribution of services to national economic policy and development and to the refinement of economic theory. Not that all services were overlooked; some non-productive services were valued for their welfare contribution to the socialist mode of production. This made it necessary to devote some resources to services such as health, science, education, sport and

Table 1.4 Service employment[a] as a percentage of total employment, former centrally planned economies of Eastern Europe, 1975–1985

Year	Bulgaria	Hungary	GDR	Poland	Romania	Czechoslovakia	USSR
1975	38.3	41.8	45.6	38.9	31.3	45.7	48.6
1980	40.4	44.6	46.4	39.4	34.7	48.0	50.3
1985	41.8	46.0	46.4	40.9	34.0	48.9	51.2

[a] Comprising employment in material services (construction, transport and communication) and non-material services (trade, utilities, science, education, cultures, art, health, sport, tourism, finance, administration).
Source: extracted from Maciejewicz and Monkiewicz, 1989, 395, table 5

social security. On strict economic policy grounds, non-material services should be starved of resources – which must be concentrated on material and productive activities in agriculture and manufacturing. But political considerations have dictated at least some support for non-material service activities.

As a result, the proportion of services in total employment in the centrally planned economies has been increasing (table 1.4). The levels achieved lie somewhere between those for developed market economies and those for developing countries. The range of shares for 1985 is spread between Romania with 34 per cent and the USSR with more than 51 per cent employed in services. If the countries listed in table 1.4 are to participate effectively at a European or global scale along more open or flexible market principles (as seems likely at the time of writing) they will not only need to narrow the 'services gap' but also need to expand to a significant degree some non-material services such as finance and insurance which are woefully under-represented (1.5 per cent in Poland, but the average was around 0.5 per cent in 1985 for the countries shown in table 1.4). A final observation is the very slow rate of expansion in services employment between 1975 and 1985 even by comparison with developing countries (see table 1.3). An economic environment that has nurtured and perpetuated inefficient and outdated production and distribution systems has not stimulated (or indeed provided any need for) the development of the business, professional and other services that are a function of the competitiveness within and amongst developed market economies.

Explanations for Growth

This brief empirical sojourn has demonstrated that, although there are differences in the level of service employment between countries according to their level of economic development or political economy, there has generally been substantial growth during the last 25 years. What are the factors that have contributed to a worldwide process that has yet to complete its full course? The first thing to note is that the acceleration in the development and diversification of service industries in the second half of the twentieth century took place against a background in which services, in the day-to-day operation of economies during the Industrial Revolution, were overshadowed by the very visible and physical impact of manufacturing production upon towns, cities and regions (Elfring, 1989a). Even before the Industrial Revolution the very earliest societies depended for their survival and identity on codes, ethics, priorities and trading arrangements that were organized by individuals with the requisite specialist skills, knowledge or status (Riddle, 1986). They not only helped to provide the structures that bound particular societies together but also facilitated the trade that was such a vital part of their economic survival and development.

Trade in goods and in services has continued to be vital to smoothing the spatial differences in the endowments of places at all scales of analysis: raw materials are not available everywhere in the same quantity and quality; environments attractive to tourism are not found everywhere; manpower with requisite skills is more highly concentrated in some areas than others; and specialist knowledge or the availability of finance for investment is more readily accessible in certain financial centres. Thus, the service that is provided as a result of trade has always been a necessity; its range and capacity before the Industrial Revolution was enhanced by the expansion of shipping routes following improvements in ship construction, and by the ability of national governments to maintain sufficient political stability to allow domestic and international trade to flourish. The British domestic market during the period of the Roman Empire, for example, was able to develop as a result of the growth of central government and the abolition of local customs tariffs (Riddle, 1986). Later expansion of colonial power by Dutch, French or British interests provided a stable framework for flows of goods and services between the colonies; many professional, financial, insurance and retail services were supplied by the colonial power, which also established the foundations of an educational system or invested in the public utilities infrastructure such as water supply systems or railway networks. As the Industrial Revolution gained momentum it encouraged greater specialization of labour inputs, allowed reductions in working hours and increased disposable incomes; the result was an increase in the demand for leisure services of the kind provided by the seaside resorts and health spas that grew quickly along the shorelines of many European countries during the nineteenth century, especially after extensions to railway systems helped make them accessible for day excursions as well as longer visits. Amusement arcades, hotels, ice-cream vendors and fun-fares were just some of the additional service activities stimulated by the economic progress that accompanied the Industrial Revolution.

Such historical antecedents for the range of contemporary services are clearly important. But it is also necessary to refer to theories of service industry growth as an aid to understanding why services have been hailed as some kind of 'new' discovery in recent decades. There is a family of general theories in which the rise of services is seen as a stage in the long-term transformation of economies. Examples of these 'stage' theories include the three-sector model (Fisher, 1935; Fourastie, 1949), the emergence of a post-industrial economy (Bell, 1973; Hirschorn, 1988), deindustrialization (Blackaby, 1978) or the transition to an information economy (Porat, 1977; Hepworth, 1989). According to these theories the service industries herald the arrival of the final stage, preceded at earlier stages by agriculture and manufacturing (Fisher); by substitution of white-collar for blue-collar occupations and the trend towards upskilling and upgrading the labour force as the key resource (Bell); or by access to, and availability of, information as the principal factor of production in place of raw material or labour (Porat). A large part of the information economy is occupied by services, although by no means exclusively (Hepworth, 1989). The service and information economies tend to be intertwined (Ochel and Wegner, 1987); the factor which binds them together, and which also explains the way in which the contribution of services to economic development and to the world economy has changed, is the revolution in information technology (IT) and communications technology which has largely occurred since the convergence of computing and telecommunications from the mid 1970s (Dicken, 1992).

The role of IT is a pervasive theme in the emerging geography of the global service economy, and it will be discussed in more detail in subsequent chapters. Suffice it to note here that IT has not necessarily coincided with a new stage in economic development in which services are central; rather it has encouraged more rapid diversification of the services available, the way that they are used in the production of goods and services, and the way in which services are used directly in relation to final demand, e.g. by using video recorders to watch movies at home rather than by making a visit to a cinema in order to be entertained. Ochel and Wegner (1987) suggest that IT has transformed economies in four main ways: what is produced (or the product mix) has been altered such that there is growing complementarity between goods and services, the development of new services and greater product differentiation (rather than mass production); markets have been changed to embrace more internationalization and growing tradability of services (see chapters 2 and 3); the location of the production of services has been modified, again including internationalization; and there has been a transformation of production processes.

The three-sector/stage theory is still widely used (sometimes with the addition of a fourth (quaternary) and a fifth (quinary) sector: Gottman, 1970), but all it really does is to describe a process of economic and/or social change that results in the services becoming the largest single group of economic activity or employment by occupation. However, this does not of course provide an explanation for the growth of service activities. One reason for this is that stage theories do not adequately disaggregate services; they are very heterogeneous and can be placed in groups that possess very different characteristics. It then becomes possible to develop some explanations for the growth of services although there is invariably disagreement about the relative importance of the factors involved. Another reason for the poor explanatory record of stage theories is that distinctions between the three major economic sectors (primary, secondary, tertiary) are increasingly difficult to substantiate. As national and global economies become increasingly intertwined, primary, manufacturing and service sectors have become more indeterminate, i.e. services have become a more integral part of all economies, especially developed market economies. This has been assisted by the emergence during the 1980s of transnational service corporations (Clairmonte and Cavanagh, 1984; UNCTC, 1990).

The stage theory of economic growth is one of a group of theories that are classified as based on production factors (Riddle, 1986). Technological progress has made increasing inroads on the production of services and has introduced a stage in the economic development process that involves a shift of emphasis from fixed capital formation in manufacturing plant and infrastructures to fixed capital formation in business services. Another theory based on production factors is the principle of comparative advantage. Traditionally this has been used to explain patterns of trade in goods, but Sapir and Lutz (1981), following Katouzian's (1970) initial suggestion, show how differences between countries in infrastructure capital and in human resources (skills, education) can be used to account for variations in the patterns of trade in services (see chapter 3).

Demand (consumption) factors can also be used to explain the expansion of service activities. Firstly, as disposable incomes rise the demand for certain services, e.g. leisure, tourism, consumer durables, private health care, also increases. Services are income elastic: they are affected by changes in tastes and priorities and by changes in price. There is disagreement about the strength of the relationship between demand for services and the degree of discretion which final consumers can exert (see for example Fuchs, 1968; Kindleberger, 1958). Gershuny and Miles (1983) use data from several European countries to show that the real consumption of privately purchased services, for example, was in decline during the 1960s and 1970s. A possible reason is that as incomes rise, the prices of services also increase rather than the quantity purchased (Kravis and Lipsey, 1983). Although the expanded provision of social and community services (see table 1.2) has increased employment in developed market economies, Gershuny and Miles (1983) conclude that the growth of service activities is not very well explained by above-average growth in the demand for services from final consumers.

The accelerating demand for producer (or intermediate) services provides a partial explanation for service sector expansion (Elfring,

1989a; 1989c; Stanback et al., 1981; Greenfield, 1966). These are the services used by primary, secondary and tertiary activities in the process of production (research and development, design, marketing and distribution, for example) of a good or service rather than by final consumers. Divisions of labour and specialization within the production process have encouraged the demand for producer services, aided by the restructuring of production within enterprises and increased externalization of functions such as advertising, security, design, packaging and computer programming. But we have seen earlier (table 1.2) that producer services have not been the principal source of service employment growth, largely because they comprise a relatively modest proportion of total service employment (even though they may account for more than one-third of service output by value). Although the demand for producer services may not be a major factor in the shift to service employment, we shall see later that they are playing a big part in the new international division of labour and in global trade in services (see chapters 2 and 3).

Another demand-related explanation for service employment growth is the belief that labour productivity improvements have been lower for services than for manufacturing. Therefore, even if demand is spread uniformly across all sectors of an economy, services will require an ever larger share of the labour force. Labour productivity, measured as real GDP per employed person, in developed market economies is consistently lower for private services than for manufacturing (table 1.5). Only West Germany has achieved a rate of increase in service sector labour productivity which is very similar to that for manufacturing. A recent study of the productivity of UK services (Millward, 1988, 274) concluded that productivity growth in the service sector over the period 1973–85 'has been low and distinctly inferior to manufacturing'; furthermore it has been consistently low since at least the mid nineteenth century. It remains the case, however, that the experience of individual countries with respect to service productivity (defined as the ratio of economic outputs to inputs) is highly variable and therefore difficult to generalize (Ochel and Wegner, 1987). It is not possible to contemplate increases in productivity in all the services; personal services such as hairdressing operate on a client/hairdresser ratio that has remained more or less unchanged throughout this century. But other services have been able to achieve large increases

Table 1.5 Annual average percentage changes in real GDP, civilian employment and labour productivity, selected OECD countries, 1973–1984

Country	Real GDP			Civilian employment			Labour productivity[b]		
	Private services	Govt services	Total[a]	Private services	Govt services	Total[a]	Private services	Manufacturing	Total[a]
Belgium	1.8	2.3	1.7	1.1	2.5	-0.3	0.4	4.6	2.0
France	3.2	1.8	2.2	1.7	1.4	0.1	1.5	3.2	2.1
Italy	2.8	1.6	2.0	2.6	2.1	0.6	0.2	2.3	1.4
Sweden	1.9	3.0	1.7	0.7	3.7	0.7	1.2	2.2	1.0
UK	2.2	1.1	1.0	1.2	0.6	-0.4	1.0	1.9	1.4
USA	3.0	1.1	1.8	2.7	1.0	1.5	0.3	1.6	0.3
West Germany	3.0	2.1	1.7	0.4	1.6	-0.5	2.6	2.8	2.2

[a] Includes manufacturing, others.
[b] 1973–83.

Source: OECD National Accounts, vol. II, 1972–84, after Ochel and Wegner, 1987, 31, table 3.1

in productivity as a consequence of advances in technology. Baumol (1967) has suggested that the same number of musicians are needed to play a Beethoven quartet in the late twentieth century as in the eighteenth century, and therefore productivity has not changed. Yet it is not difficult to argue that technological advances in the recording, reproduction and transmission of music have made it possible for an almost limitless number of people to listen to the music; in these terms, the productivity of the musicians has certainly improved. Distribution and communication services have been able to improve productivity by large margins as a result of technology (see for example Baumol, 1987).

But we have seen how employment in services has been expanding even if some sectors have witnessed improvements in productivity. This paradox arises from the role of the 'information economy' (Porat, 1977; see also Castells, 1989) which is a *new* phase of economic development, wherein the production of information goods and services dominates wealth and job creation with computers and telecommunications providing the technological potential for product and process innovation' (Hepworth, 1989, 7). As information processing activities have expanded, productivity growth has not been able to keep pace. Information is used to improve the productivity of manufacturing, extractive and service activities but the management, acquisition and interpretation of that information are labour intensive, even though powerful information processing technology is now available. The production and assimilation of information generates a requirement for yet more information to assist with the analysis and interpretation, for example, of that which is initially available. In this way information occupations such as scientists and technicians, educators, consultants and market researchers now comprise 30–40 per cent of the labour force in advanced economies. Castells (1989, 136, after de Bandt, 1985) suggests therefore that: 'Behind the expansion of the service sector, directly in terms of employment, and indirectly in terms of its effects on output, lies the development of the information economy.'

The debate about the relationship between service productivity and its employment effects has also endured because of the persistent problems of how actually to measure service output (see for example Smith, 1972). On balance, however, Gershuny and

Miles (1983, 40) conclude that the productivity rate of the service sector is lower and give considerable weight to 'explanations which relate the growth of employment in the services to differential productivity growth'.

Riddle (1986) cites two other demand-related factors in service sector expansion. The first is the level of urbanization which Singelmann (1978) hypothesized would cause expansion of the service sector. However, only upper-middle-income countries in Riddle's fourfold classification of development (low income, lower middle income, upper middle income and industrial) revealed a significant correlation between percentage population in urban areas and services as a proportion of GDP in 1981 (the correlation was not statistically significant for 1977). It may even be the case that urbanization is negatively correlated with services as a proportion of GDP if urban centres are smaller than 1 million population. The second link between demand and service sector growth arises from international trade or export led growth. One of the most striking developments of the 1970s and 1980s has been the growing interdependence of nations within the global economy (see for example Dicken, 1992; Knox and Agnew, 1989), with trade becoming as important for national economic well-being as domestic production and consumption. Since the Industrial Revolution it has been assumed that such trade is dominated by flows of raw materials and manufactured goods and is dictated by the needs of final consumer markets; but trade generated by the requirements of intermediate markets (i.e. the producers of goods and services that have adopted global location strategies: see for example Dicken, 1992) must now be included in the total. Furthermore, international trade in services not only binds together the developed market economies; it also includes low-income countries (UNCTAD, 1989a; Riddle, 1986). It has also been recognized in recent years that institutional factors such as changing product and market strategies, corporate structure and public policy are also important variables affecting the growth of services at national and international level (see for example Shelp, 1981; Singelmann, 1978).

The response to the fast changing dynamics of the demand for services has been characterized by some as flexible specialization (Piore and Sabel, 1984; Cohen and Zysman, 1987; Gertler, 1988). Producers of goods and services must adopt a strategy that aims to achieve permanent innovation and adaptation. The flexibility is made possible by attracting and training broadly skilled labour, by developing networks of interfirm cooperation and by using the flexibility of technology to its full potential. Thus, with the same stock of capital and labour, firms can guarantee that a wide range of services is developed, produced and distributed. Furthermore, flexibility enables services with a limited production life to become practical from an economic point of view; it also enables firms to adopt location patterns that are more dispersed, with less centralized organizational control. The networks between firms allow specialized producers to function within highly segregated production systems that further encourage adaptation and innovation amongst large and small firms alike. Whereas firms in an industrial economy are largely concerned to reduce and control costs, to vertically integrate production and to view labour as a cost and a source of physical effort, in the post-industrial economy firms are conscious of the need to reduce exposure (just-in-time systems are more common), to control quality, to vertically segregate production (including subcontracting) and to view labour as a capital asset which contributes adaptability rather than effort (Hirschorn, 1988).

It will be apparent that the theories available to us in our quest for an understanding of the growth of services or of the relationship between services and economic development leave much scope for uncertainty. The stage theory of economic development and the place of services within it is hard to substantiate; indeed in some countries services have always been more prominent than manufacturing, while in others services are growing much faster than manufacturing even though they are only just beginning to move away from a dependence on their primary resource base. Some of the theories are useful for understanding national trends in economic development; others are more relevant to international trends and the contribution of services to them. Noyelle and Stanback (1988) emphasize the need to complement these theories with a closer examination of what is being produced and how production takes place. They list five transformations that are reinforcing the shift to services: the rise of large corporations; technological change; increases in market size and product differentiation; the development of new consumer markets; and the growing influence of government and non-profit organizations. Some of these have

Table 1.6 Determinants of total economic activity, countries grouped by development category, 1981

| | Development category | | | | | | | |
| | Low income | | Lower middle income | | Upper middle income | | Industrial | |
Factor[a]	Dom.[b]	Int.[b]	Dom.	Int.	Dom.	Int.	Dom.	Int.
1	Growth	Trade	Service employment	Service employment	Quality of life	Trade	Personal income	Trade
2	Employment	Growth	Growth	Personal income	Growth	Personal income	Investment	Personal income
3	Quality of life	Employment	Quality of life	Growth	Service employment	Growth	Services GDP	Manufacturing GDP
4	External debt	Personal income	External debt	External debt	External debt	External debt	Growth	Growth
5	Personal income	–	Services GDP	Services GDP	Services GDP	Manufacturing GDP	–	–
6	–	–	Inflation	Inflation	Manufacturing employment	Manufacturing employment	–	–

[a] Each factor comprises a number of variables which are positively or negatively loaded (in the range +1.0 to −1.0).
[b] Domestic and international economic data.
Source: extracted from Riddle, 1986, 54–63, tables 2.9–2.12

already been mentioned earlier in this chapter and will be returned to later in this book.

Understatement of Role of Services

The debates about the relative merits of the various theories and explanations for the expansion of services will clearly continue. One consistent feature is the way in which they highlight a long-standing understatement of the role of the service sector in national and global economic development. In relation to the global perspective, Riddle (1986) has underlined the position of services by examining the independent determinants of total economic activity in four groups of countries (table 1.6). The variables in the factor analysis used to identify the groups (in order of their loadings) shown in table 1.6 include (for each country): rate of urbanization, economic growth, financial indicators (such as the inflation rate, domestic savings and investment), sectoral percentages for GDP and employment, debt variables, trade variables and quality of life indicators. This analysis clearly shows that, with the exception of low-income countries, services are prominent determinants for total economic activity in countries at all levels of development, and that the variability between the economic performances of countries is as much to do with factors related to services as with factors related to manufacturing or primary sector activities. Table 1.6 also confirms that the export of services is not just an activity characteristic of developed market economies: it is also important for developing economies; especially the lower-middle-income countries where service exports are closely associated with economic growth (Riddle, 1986). Clearly, as Elfring (1989a) has observed, the expansion of services is intertwined, in a matrix of growing complexity, with other parts of the economy (see also Barcet, 1988). Increasing specialization in production (as well as more complexity) has stimulated the rapid expansion of producer services, while changes in household structure and greater participation in the labour force by females have raised the demand for personal services.

To summarize, the combination of demand-side and supply-side factors explains the increased share of services in GNP in six ways (Dunning, 1989). The first is the growth in demand for discretionary consumer services following the growth of per capita output. The second is the steady emergence of a role for producer or intermediate services in the value added to a good or service. The third is the trend towards external purchase of services by non-service firms. The fourth is the significance attached by producers to effective marketing, distribution, after-sales maintenance and servicing of their products, and by governments to investment in social and community services or infrastructure such as telecommunications. The fifth is the more specialized or sophisticated demands of contemporary societies for legal, insurance, transport, banking, entertainment and financial services. The sixth is the ability of service producers to create new products and new markets in areas such as securities, junk bonds, reinsurance, debt swaps, Euro-markets, value-added services and data transmission and manipulation.

The Tradability of Services

Tradable and Non-Tradable Services

In order for services to become prominent actors in the world economy they must possess attributes that enable them to be exchanged across geographical space. This is essentially dependent on the extent to which they are tradable, i.e. they can be offered on the open market and purchased by customers several thousand miles away from the suppliers, or they can be traded within organizations but between establishments located in several different countries. Not all services are tradable; some can only be directly consumed or purchased from 'fixed' points of supply. Examples include consumer services such as clothes retailing and fast-food restaurants, warehousing, transport services, hotels, public utilities, and personal and household services. Services of this kind require direct interaction between supplier and purchaser; traded services may involve a similar interaction but will often be provided by using some kind of intermediary such as telecommunications or courier services. Hirsch (1986) has used the expression 'simultaneity factor' to explain the difference between tradable and non-tradable services: the lower the proportion of their total costs incurred by the producer and the user during their interaction, the greater the tradability of a particular service.

Traded services are therefore those supplying 'business and government organizations, rather than private individuals . . . and may even be indirectly tradable through their contribution to the competitiveness of other sectors of the economy' (Marshall et al.,

1988, 13). These are the producer (or intermediate) services, which can be grouped into, firstly, information processing services such as banking, insurance, marketing, accountancy, property management, advertising, codifiable information and technology, and a variety of property rights such as patents, tape recordings and architectural drawings (Dunning, 1989); secondly, goods-related services such as distribution, transport management, infrastructure maintenance and installation, repair and maintenance of communications equipment; and thirdly, personnel support services such as welfare, catering, personal travel and accommodation (Marshall, Damesick and Wood, 1987). The empirical evidence outlined in chapter 1 showed that in most economies the employment in services has been growing fastest in producer services even though they do not form the largest part of the service sector. As we shall see in chapter 3, it is these services that have been responsible for the growing involvement of services in the world economy as well as within national economies.

The main aim of this chapter is to examine some of the factors that have permitted or encouraged worldwide transactions in services, whether between or within organizations, as well as the internationalization of the sources of supply.

Information Technology and the Tradability of Services

The value of cross-border trading in shares by the US stock exchanges more than doubled between 1984 and 1986 and tripled during the same period on the Japanese exchanges (*Financial Times*, 1987). The lifting of controls on the operation of national financial markets and on movements of capital was an influential factor in this growth, but the increasing efficiency of international communications and the associated information technology (IT) revolution were probably more significant. Information technology is 'pervading virtually all forms of human endeavour: work, education and leisure, communication, production, distribution and marketing and the time scheduling of these. It is changing the scale and content of information networks, the interdependence of organizations, and how as well as where we live, work, shop, learn, communicate and play' (Brotchie, Hall and Newton, 1987, xv; see also Brunn and

Leinbach, 1991). Information technology enhances to a very significant extent the number of people that can engage in the organization of knowledge, the speed and the volume of knowledge that can be processed and exchanged, the significance of location for interaction potential, and therefore the dynamics and diversification of economic activity. It is therefore 'widely held to be of crucial importance to development and economic success both in individual enterprises and in nations' (House of Commons Trade and Industry Committee, 1988, after Hepworth, 1989, 1).

Information comprises factual data such as statistics, the results of experiments, bibliographical data, business documents, the characteristics of processes, or forecasts and models (both subjective and objective) of trends and outcomes based on available data or experimental results. It is transient and of no intrinsic value until exchanged with others who attach value to or gain from its receipt; it is held by a person or persons; it can be transmitted to other persons in the form of a message; it requires a person or persons to receive the message (or a machine programmed to receive and process the message using specified procedures); and, finally, the message has to be comprehended. This is a rather wide definition but it helps to explain why so much modern economic activity is derived from the origination, processing and dissemination of information (Gibb, 1982). It has also spawned wide ranging studies of the information society (Beniger, 1986), the information economy (Porat, 1977) and the geography of the information economy (Hepworth, 1989). According to Hepworth (1989, 7), 'the information economy is a new phase of economic development, wherein the production of information, goods and services dominates wealth and job creation with computers and telecommunications providing the technological potential for product and process innovation.' The requirement for information exchange, for example to advance individual or organizational interests, has always been present. What IT has done is to greatly enhance the way in which information is handled by persons or machines, to revolutionize the capacity and speed of the 'channels' used to move it between origins and destinations, and to diversify the methods and opportunities for delivering services. All have been keys to advancing competitiveness in services – especially those that are tradable. Barriers to market entry have also been lowered by IT.

The two main advances since the 1960s have been in microelectronics and in telecommunications. In microelectronics there has been a staggering increase in the number of memory bits (used in logical processing) that can be stored on one silicon chip. The chip is the key component of everything from hand-held calculators through desktop personal computers to large mainframe computers, and is so reliable that the number of failures per unit of time has been reduced by a factor of at least 10 million since the 1950s (Danzin, 1983). These and other advances have greatly reduced the size of the computer hardware required to perform a given task (many powerful machines are easily portable) as well as brought down prices and made the technology more accessible to a much wider range of users. A fifth generation of computers capable of artificial intelligence is now being developed, with Japan and the United States vying to be the first to produce such machines.

But computers, unless used for totally self-contained purposes, are of relatively limited value if they cannot communicate with each other. The same is true for persons: information has limited value unless it can be passed on or received using a medium that overcomes the distance separating the communicators. Physical methods of transmission such as postal services have long been available, but the speed, capacity and cost of these services have restricted the intensity and quality of the information that can be transmitted. As an electronic channel for the transfer of voice or data between any two points (as well as mass diffusion of information through broadcasting) telecommunications has revolutionized information exchange. For almost a century prior to the late 1940s telecommunications (the telephone and the telegraph) depended on analogue transmission. This gives direct and continuous wavelike representations of voice or audio signals; the speed of transmission is therefore slow and the kind of information transmittable is limited. Digital transmission involves the transfer of information which has been converted to binary form, leading to a marked increase in speed and capacity as well as in the diversity of information that can be handled. But this has also required an improved medium for transmission since the traditional copper-wire-based telephone networks are unreliable, slow and restricted in capacity. Fibre-optic cables (which use modulated light wave signals generated by lasers) have overcome these problems and have further underpinned the convergence of telecommunications and computing.

There are two characteristics arising from this convergence which are significant for the tradability of services. Firstly, the convergence has allowed services to be produced and made available using a common storage and transmission medium. Secondly, it has enhanced connectivity and interoperability in that a large number of users, whether persons or machines, can process or otherwise share resources across the boundaries of telecommunications networks (Pipe, 1989). At the international level, networks are not just based on transcontinental or transoceanic cables but include satellite systems. For example, the International Telecommunications Satellite Consortium (Intelsat) launched its third satellite in 1968 with a capacity of 1500 telephone circuits and four channels for television. Its fourth satellite, launched in 1989, has a capacity of 120,000 bidirectional circuits for transcontinental telecommunication. The transoceanic cable between Britain and the US completed in 1987 (TAT-8) has 40,000 circuits; a second cable (PTAT) which became operational in 1989 has a capacity of 80,000 circuits. Clearly, the capacity (as well as the speed) of the international telecommunications infrastructure has been growing rapidly, especially transoceanic links (figure 2.1) (see for example Janelle, 1991;

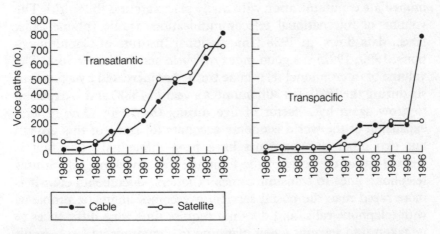

Figure 2.1   Growth (actual and estimated) in capacity of transoceanic cable and satellite telecommunications systems, 1986–1996
Source: International Institute of Communications, 1991

Langdale 1991). But so too has the cost of staying in touch with the latest technology. Financial services have been leading the way in the adoption of IT: by 1988 each of the largest US banks was spending $200m a year on IT, while a large UK clearing bank such as Barclays was spending $300m a year (McKinsey and Company, 1987). The seven largest US banks will probably have spent 40 per cent of total US bank spending on information technology in 1990 (estimated $6.5b). IT equipment has a relatively short lifespan because new innovations invariably reduce the effectiveness of existing equipment and software, and this ensures that for many services the commitment to IT will remain a major drain on resources. They will continue to invest because IT is a crucial part of maintaining competitive advantage.

From the perspective of this book, the key feature of the IT revolution is the way in which it has virtually eliminated the effect of distance on the time required to communicate an item of information between interacting locations. It has effectively eliminated the role of national borders as physical or legal barriers to service transactions. The Intelsat communications satellite, for example, is owned by a consortium of some 110 nations and links 170 countries by earth–satellite–earth signals that take less than a quarter of a second to transmit. Thus both developed and developing economies have the capability of reliable and effectively immediate communication with many places around the world. The volume of international telecommunications traffic (phone calls, faxes, data flows) in 1988 (International Institute of Communications, 1990; 1991) is a good index of world activity in services. The volume of international telephone traffic has increased by a factor of six during the 1980s (to 30b minutes a year in 1988) and is expected to grow again by a factor of five during the 1990s. The general expansion of the world economy accounts for some of this growth but two other developments have been influential. The first is technology in the form of the facsimile (fax) machine, which uses telephone lines to transmit copies of letters, documents etc.; it is more rapid than the postal service, overcomes linguistic problems with telephone calls, and does not require time zone differences to be taken into account when planning to communicate. As a result some 60 per cent of transpacific telephone traffic involves the use of facsimile machines. The second factor underlying the boom in international telephone traffic is the development by multinational service corporations of complex telecommunications networks that allow extensive and intensive intra-organization communications and sophisticated relationships with suppliers and customers. These networks are part of the value of a service to clients and the source of competitive advantage of one service corporation over another. According to Dixon and Staple (1991), the study of who is tele-communicating with whom gives a new perspective on the patterns of power and influence in the world. Indeed, 'telegeography' is more revealing than traditional maps because telecommunications has become involved with almost every branch of human activity. Most service activities are involved with the telecommunications network at least once; some, such as those engaged in the trading of foreign exchange, are totally dependent on the network. Worldwide there are more than 1 billion telephones linked by a network of ground cables and satellite links – a sort of nervous system for the global economic and social body.

The origins and destinations of international telephone traffic are unevenly distributed around the globe (table 2.1). Outgoing calls from the United States totalled 5.3b minutes in 1990; the equivalent figure for Malaysia was 0.08b minutes. International calls from Portugal totalled 0.13bn minutes and from South Korea 0.19bn minutes. The unequal distribution of telecommunications services and traffic is underlined by the example of India, which not only lacks international connections but also has fewer telephones than London. Apart from differences in telecommunications infrastructures and the international orientation of domestic economic activity, the cost of international telephone calls is an important source of variation in volume. The prices for international calls from Italy are 40 per cent higher than elsewhere in Europe; Italy made 55 per cent fewer outgoing calls than the UK in 1990. Japan also has a surprisingly low volume of outgoing international calls, and again this reflects high prices. However, as prices fall in countries that are major destinations (perhaps because of the removal of telephone monopolies, as in the UK) high-cost countries will have to follow suit. Currently, telephone traffic out of Japan is growing at twice the global average as prices fall and Japanese manufacturing and service business continues to extend its global presence.

All these interactions are often described as transborder data

Table 2.1   International telephone traffic, selected countries, 1990

Country	Telecommunications traffic (million minutes)		
	Outgoing	Incoming	Balance
Australia	518	398	−120
Japan	764	732	−32
South Korea	188	350	162
Malaysia	80	100	20
Taiwan	212	302	90
Austria	476	487	11
Belgium	731	755	24
Canada	565	358	−207
Denmark	362	343	−19
Finland	186	213	27
France	1921	2190	269
Germany	2833	2369	−464
Ireland	75	122	47
Italy	908	1161	253
Israel	118	202	84
Luxembourg	151	83	−68
Netherlands	905	852	−53
Norway	281	277	−4
Portugal	126	270	144
Spain	611	653	42
Switzerland	1356	1016	−340
Turkey	159	441	282
UK	2253	2330	77
US	5265	2604	−2661
Canada	565	358	−207

Source: International Institute of Communications, 1991

flows. Not only do they facilitate the tradability and ultimately the internationalization of service activities, but they have become a service in themselves. Transborder data flows can substitute completely for the physical movement of goods, persons or capital when items such as architectural drawings, design proposals, legal contracts, share prices, currency movements, export manifests, loan agreements, and insurance and banking documents can all be transmitted electronically and safely. Transborder data flows are probably underestimated (and increasingly so) because no payments may be involved to a legal entity outside a country's borders (UNCTAD, 1989a) or because they are part of the flow of information within multinational companies, some of which already have established (or are in the process of so doing) private telecommunications networks in connection with the control and management of decentralized corporate systems. The international IT networks created to facilitate transborder data flows also provide channels for delivering services to markets and add to competitiveness in trade in goods.

In addition to accelerating the flow of information and increasing the range of markets, IT has also created opportunities for a whole new range of services that could not be competitive or cost effective in other circumstances (see for example Howells, 1988; Porat, 1977). On-line information services such as Quotron (financial information), Equifax (financial control) and many others in the United States all increased their subscribers and revenues by large margins in 1985 (Howells, 1988, table 4.1) – more than 100 per cent for customers in some cases, and between 4 and 70 per cent for revenue. In 1984 Reuters, which provides a worldwide real time information service on items such as share and commodity prices and currency quotations in different markets, had 65,000 subscribers (an increase of 21.6 per cent for the year) and a revenue of $310m (up 24 per cent on the year) (Howells, 1988). Value-added networks (VANs) – publicly accessible services based on reformatting, storage and retrieval of information in a way that makes it 'valuable' to subscribers – tripled their annual revenues in Western Europe between 1984 ($270m) and 1986 ($900m) (Frost and Sullivan Inc., 1985). A measure of the rapid emergence of information services companies is the growth of the revenue from this activity for the world's twenty leading companies between 1986 and 1989, both in real terms and as a share of their total revenue (table 2.2). A number of firms such as American Express or Ernst & Young did not provide information services ι 1986 but within three years were obtaining a significant proportion of their overall revenues from the provision of these services. Table 2.2 also demonstrates the way in which diversification has characterized the development of international service firms: only six firms in the top twenty rely on information services for more than 80 per cent of total revenues. By providing a more comprehensive service to clients

Table 2.2 Leading information service companies in the world, 1986 and 1989

Firm[a]	Home country	1986 Information services revenue[b]	1986 % of total revenue[b]	1989 Information services revenue	1989 % of total revenue
Electronic Data Systems Corp.	US	0.0	0.0	2447.9	45.3
Automatic Data Processing Inc.	US	1298.1	100.0	1689.5	100.0
TRW Inc.	US	1450.0	24.0	1565.0	21.3
Computer Science Corp.	US	977.7	100.0	1442.8	100.0
Digital Equipment Corp.	US	0.0	0.0	1386.7	10.7
Andersen Consulting	US	0.0	0.0	1225.7	85.0
International Business Machines	US	300.0	0.6	1200.0	1.9
Cap Gemini Sogeti	France	0.0	0.0	1103.4	100.0
NTT Data Communications Corp.	Japan	577.6	1.8	898.7	39.9
Unisys Corp.	US	0.0	0.0	825.0	8.2
Black & Decker Corp.	US	0.0	0.0	687.6	21.6
American Express Co.	US	0.0	0.0	660.0	2.6
General Electric	US	550.0	1.5	550.0	1.0
Martin Marietta Corp.	US	0.0	0.0	502.2	8.7
Ernst & Young	US	0.0	0.0	450.0	11.5
NCR Corp.	US	350.0	7.2	425.0	7.1
STC PLC	UK	0.0	0.0	424.6	10.0
Sligos	France	0.0	0.0	385.5	96.2
SD-Scicon PLC	UK	0.0	0.0	381.3	82.3
British Telecommunications PLC	UK	0.0	0.0	360.2	1.8

[a] Ranked by total information services revenue, 1989 (millions of dollars).
[b] Value 0.0 denotes firms not operating in information services in 1986.

Source: UNCTC, 1991, after Datamation, 1990 and 1987

which draws upon their particular knowledge in core activities such as telecommunications and management consulting, firms such as STC or Ernst & Young are hoping to retain long-standing business relationships that are increasingly threatened by specialist suppliers.

Capital markets have been revolutionized by network technology that allows 24 hour dealing in foreign exchange, international securities, futures and other equities that have been developed since electronic trading became a major force behind the internationalization of equities trading (figure 2.2). The US National Association of Securities Dealers Automated Quotation (NASDAQ) system, which was established in the early 1970s, has recently been linked with the London Stock Exchange Automated Quotations (SEAQ) system to provide a transatlantic trading service in equities that will include monitoring, surveillance and computer-based quotation of prices. This network will give wider decentralized access to trading information that previously could only be obtained quickly by traders on the spot at the London or New York Stock Exchanges. These services are largely used by businesses, but services such as Oracle and Prestel (UK), Minitel (France) and Viatel (Australia) are also targeted at private consumers and offer home banking, shopping, weather, travel timetables and similar information. While service businesses have been quick to appreciate the benefits of access to information services using IT, private consumers have been much more reticent about substituting face-to-face contact, or travel to make a purchase from a retail outlet, with decisions made entirely on the basis of voices and images conveyed by an electronic medium.

Information technology has been responsible for convergence in the way that globalization has been undertaken by different types of services (Vandermerwe and Chadwick, 1989). Self-diagnostic systems can be built in to computers, reducing the need for regular or emergency on-site checks by maintenance staff or programmers. Such visits need only take place when the diagnostic system (which may include some problem-correcting procedures) indicates that input is required. Even then this input may be achieved through a network of terminals from any part of the globe, and individuals will not need to travel to a malfunctioning computer system to undertake repairs. Loan finance on a large scale is now arranged from a limited number of locations around the globe using telecommunications

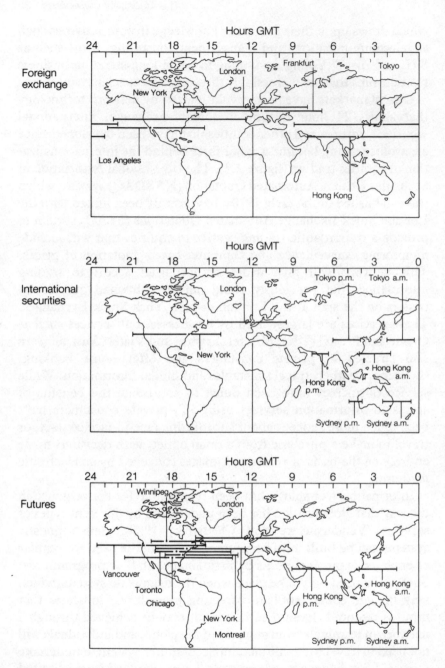

Figure 2.2 Global trading of three financial markets
Source: after Clarke, 1986

networks (both intrafirm and public access) to conduct transactions with clients and between the cities at the heart of the global financial system (see chapter 4). Electronic machines are now used to sell services such as travel or car insurance at international airports or in large shopping malls. Physical presence in the form of a branch office is not necessary for insurance companies to globalize their activities: electronic networks and machines give them representation in as many host countries as will allow them to operate (see chapter 3). Innovations in the delivery of educational services using television, broadcasting and self-teaching packages for use on personal computers mean that students no longer need to congregate in one location to acquire knowledge or training; it can be delivered to them nationwide (e.g. the Open University in the UK) or worldwide while still allowing almost simultaneous acquisition and assessment of progress through a course. It is, however, extremely expensive for individual large corporations to install their own international telecommunications network. This has, for example, been recognized by British Telecom which is about to establish a global digital telecommunications network by persuading Deutsche Bundespost Telekom and Nippon Telegraph and Telephone to take major stakes in Syncordia (an American-based company) which it established in 1990. The integrated sets of voice, data and video communications circuits that form global networks, already used by IBM, Ford and Shell, are one of the fastest growing and most lucrative areas of telecommunications. Service and manufacturing multinationals will be able to turn over the management of their networks to Syncordia, which will provide circuit ordering, repairs, customer assistance and billing 24 hours a day for a network linking 70 of the largest financial and manufacturing centres around the world.

The demand from businesses for services using high-speed digital networks will doubtless continue, although lags in accessibility to such services between countries will persist in line with the investment priorities of suppliers and the effects of national regulations on market penetration (Pipe, 1989). This will act as a barrier to economic development in the case of countries with outdated systems and as a stimulus to those with the latest equipment and facilities (Newton, 1989). It has been estimated by Newton (1989) that the telecommunications network of the future, a broad-based digital fast packet-switched optical fibre network, may take at least twenty years to achieve for technologically advanced societies. We shall see in later chapters that this has important implications for the patterns of international trade in services, for the location decisions of the organizations involved and for global patterns of economic development.

Transport Technology and the Tradability of Services

Although IT revolutionized the market for services and their delivery, it remains difficult to deliver many of them without recourse to the movement of persons. Many services are embodied in persons, e.g. the information and skill provided by a person in order that a good or service can perform the function for which it was supplied. Much does of course depend on the kind of service. Shipping lines attempting to improve the cost effectiveness as well as the speed of handling and delivery of shipments have not found it necessary to increase the number of crew sailing between ports; rather they have adopted containerization, larger vessels for deep sea routes, roll-on/roll-off facilities for short sea routes, and high-speed transfer of containers direct from deep sea vessels to smaller coastal or river vessels (special containers can even be linked together to form units that can be pushed up rivers such as the Rhine from Rotterdam). Auditing services undertaken by accountants require the physical presence of the appropriate staff at the establishments being audited; in the case of multinational firms these will be worldwide. The installation of sophisticated computer software requires the presence of individuals familiar with its configuration and able to solve problems that emerge during commissioning and testing.

Demand from business travellers, together with travel connected with tourism (itself a major service activity), is reflected in the expansion of international airline services. The ability of airlines to meet demand as well as to encourage further growth has been made possible by improvements in flight times on long-haul routes, by increasing the range and capacity of aircraft and by using supersonic as well as subsonic speeds. However, the environmental side effects of supersonic flight (notably the sonic boom experienced at ground level) have limited routes to ocean sectors, effectively the high-

Table 2.3   International passenger destinations, New
York City airports, 1982 and 1987

Destination city	Passengers (000s)		% change
	1982	1987	1982–7
London	699	1032	47.6
Paris	343	506	47.5
Frankfurt	327	450	37.6
Rome	261	321	23.0
Amsterdam	196	268	36.8
Madrid	164	193	17.7
Shannon	148	113	−23.5
Montreal	143	262	83.2
Tokyo	141	236	67.4
Milan	125	197	57.4
Zurich	123	197	60.2
Mexico City	117	147	25.5
Brussels	110	127	15.5
Copenhagen	110	137	24.5
Rio de Janeiro	104	105	1.0
Athens	66	102	54.6
Total 16 cities	3177	4393	38.1
Total 66 cities[a]	3783	7028	85.8
% 16 of 66	84.1	62.5	

[a] Total of international destinations from New York is 66.
Source: Moss, 1989, Table 7, after International Civil Aviation
Organization, 1982, 1987

demand trans-atlantic corridor between Paris/London and New
York/Washington. While it is not possible to disaggregate air
passengers to distinguish between business and non-business
travellers, it is safe to assume that the large absolute and
proportional increases in international air passengers between New
York and sixteen cities (with 100,000 or more passengers from New
York) in 1987 are a reflection of the demand from executives
employed by multinational manufacturing and services businesses
(table 2.3). In 1987 almost one in seven international passengers
from New York travelled to London, and one in fourteen to Paris.
The concentration of international passenger flows and destinations
is related to the way in which services have expanded outside their
domestic markets and in particular the locations which they have

Table 2.4 Daily non-stop international flights in the Asia-Pacific region,
1970–1990

Pacific Rim city	Cities[a]				Flights[b]			
	1970	1980	1990	Change 1970–90	1970	1980	1990	Change 1970–90
Anchorage	1	0	1	0	1	0	1	0
Auckland	0	1	2	2	0	1	4	4
Bangkok	3	3	10	7	5	13	33	28
Guam	0	1	2	2	2	1	4	2
Hong Kong	5	9	20	15	15	32	77	62
Honolulu	3	5	7	4	6	12	17	11
Jakarta	0	2	4	4	0	17	17	17
Kuala Lumpur	2	2	3	1	8	18	21	13
Los Angeles	9	13	20	11	17	40	62	45
Manila	2	3	6	4	2	8	13	11
Melbourne	0	0	1	1	0	0	1	1
Nadi	2	0	0	−2	2	0	0	−2
Osaka	1	5	7	6	5	12	12	7
San Francisco	2	10	10	8	6	19	19	13
Seoul	1	2	15	14	3	4	35	32
Singapore	5	9	19	14	13	54	60	47
Sydney	1	2	7	6	1	2	10	9
Taipei	4	5	11	7	17	20	41	24
Tokyo	8	10	24	16	18	28	74	56
Vancouver	3	6	10	7	12	22	46	34
Total	52	88	179	127	133	303	547	414

[a] Number of cities with connections from named city.
[b] Number of international flights originating in named city.
Source: Scott and O'Connor, 1991

chosen (see chapter 4). It is notable, however, that the share of total
numbers of international passengers travelling from New York to
the sixteen destinations shown in table 2.3 has declined from 84 per
cent in 1982 to 63 per cent in 1987. The extent of this concentration
over time is demonstrated in table 2.4, which traces the growth of
daily non-stop international flights in the Asia-Pacific region
between 1970 and 1990 (Scott and O'Connor, 1991). The total
number of cities served by flights originating in the cities listed has
more than trebled between 1970 and 1990, but a large part of the
increase has focused on just four of the twenty cities listed: Tokyo
(56 additional daily non-stop flights), Singapore (47), Los Angeles

(45) and Hong Kong (62). As the figures for the number of cities served by these flights show (table 2.4), growth in connectivity involves the same centres along with Seoul. Supply and demand for international travel (business and tourism) exert a powerful influence on the patterns of growth suggested by table 2.4, although large cities do not have exclusive access to transoceanic non-stop services: some locations such as Nadi or Honolulu in the Pacific have benefited from the stop-over requirements of all but a small number of truly long-haul aircraft (Boeing 747–400s, for example) and therefore have good international connections. This has also had the effect of boosting the tourism economy of these islands (Scott and O'Connor, 1991). Although a very crude indicator, this suggests that the internationalization of services is beginning to embrace locations and countries outside the established foci such as London (Western Europe) and New York (North America).

While improvements in the speed, capacity and route coverage of air transport have been particularly significant for international business travel, within continents there is continuing scope for competition between air travel and high-speed ground transport. The potential has been demonstrated by Hall (1991a). Depending upon the kind of motive power used for high-speed trains, the break-even distance between train and air is estimated to be 530 km for a 200 km/h train service (high-speed train (HST)), 960 km for a 300 km/h (such as the Train à Grande Vitesse (TGV) in France), and 2688 km for a 500 km/h train using magnetic levitation (maglev) technology. The Japanese are proposing a 950 km/h service, also using Maglev, for which there is no break-even distance in that it is marginally faster than an air service between pairs of city centre destinations. It is likely that as European rail networks become better integrated – a process encouraged by the creation of a unified European market in 1992, the opening of the Channel (rail) Tunnel and the adoption of more standardized motive power (perhaps based on TGV technology) – many trans-European business connections will switch from air to train. If HST technology is used then London–Paris, Paris–Frankfurt and London–Brussels will be faster than by air; using TGV technology, Manchester–Paris, Lyon–Brussels and London–Frankfurt; and using Maglev, Edinburgh–Madrid, Lyon–Warsaw and London–Seville. The Community of

European Railways (twelve EC countries plus Switzerland and Austria) estimated in 1989 that rail travel will increase 400 per cent during the next 30 years, especially after 1992 (Hall, 1991a).

These developments have been sketched here not just because they offer opportunities for the expansion of service industry markets across national boundaries within Europe, but because of what they may mean for the locational pattern of the suppliers of traded services. It will be evident that if travel switches from air to rail within Europe it is likely to boost the growth potential of central and inner city locations from which most rail routes will start and finish, but it may also boost locations along the routes of high-speed services since suppliers may not need to compete in distant markets by establishing a physical presence.

Service Multinationals and the Tradability of Services

The decade of the 1980s has been one of globalization of business. Very rapid advances in information technology combined with improvements in international transport services since 1960 have been the catalyst; service enterprises are looking more towards international markets. There is already extensive literature on the theory and on the attributes of multinational enterprises (MNEs) (Buckley and Casson, 1976; 1985; Caves, 1982; Dunning, 1981; Rugman, 1979). Until comparatively recently, however, this literature has been devoted to an examination of manufacturing MNEs. This bias is explained by an emphasis on foreign direct investment (FDI) by manufacturing industries during the period after the Second World War, and the need for overseas-based firms to possess some form of 'compensatory advantage' when trying to compete with indigenous firms (Enderwick, 1989a). The policy issues arising from the development of MNEs had also been largely seen as connected with manufacturing. But in recent years the rate of FDI in manufacturing has begun slowing down (OECD, 1981), perhaps as part of a maturing process in international production (Dunning, 1981). It has become more difficult to separate goods from services in that most of the former embody non-factor services in their production and the latter require physical assets and intermediate goods for their delivery (see for example Levitan,

1985). Demand has undoubtedly been expanding for a variety of advanced services such as management consulting or commercial banking which can assist with the restructuring strategies of manufacturing MNEs. Most goods now embody some non-factor intermediate services, and most services embody some intermediate goods (Dunning, 1989; Grubel, 1987). The development of service MNEs has therefore been the focus for a number of studies undertaken during the 1980s (Dunning and Norman, 1983; 1987; Enderwick, 1989b; Schwamm and Merciai, 1985; Yannopoulos, 1983; Daniels, Thrift and Leyshon, 1989).

What are the factors prompting service enterprises to expand across borders? The motives are diverse and there is no all-embracing explanation (Dicken, 1992). Throughout the industrial-ized world, established market positions are under threat: deregula-tion, accelerating technological innovation and the rapid emergence of new competitors have undermined long-standing market boun-daries. In order to sustain growth and profit expectations it is now necessary to gain access to new markets. Services that have thrived on traditional home markets can no longer afford to do so, either because those markets are saturated and there is no scope for further product innovation or because of the effects of deregulation on the competitiveness of service firms in other geographical markets. As this competition has become more global, service firms have had to become more aggressive about international expansion strategies. As a result, there has been consolidation of European banks in recent years, for example, as they attempt to delay predatory takeovers or to deny possible bid targets to competitors eyeing the same targets. This is a simple example of the competitive strategy which a firm needs to consider, and which is especially important for such a major decision as going international. Some of the most detailed work on competitive strategies has been undertaken by Porter (1986), who suggests that there are two core strategies to consider: aiming to be the lowest-cost producer of a service (cost leadership), and attempting to be different from competitors (differentiation). A third component of the strategic decision is whether to go for a niche based on a particular product or geographical market or to operate on a more general front (focus).

Perhaps the most detailed research on the theory and practice of international production by service enterprises has been undertaken by Dunning and his colleagues (for an excellent overview see Dunning, 1989; see also Edvardsson, Edvinsson and Nyström, 1993). An MNE is 'an enterprise which owns or controls value-adding activities in ten or more countries' (Dunning, 1989, 5) and its strategic aim is to produce and deliver services (or goods) more successfully than its competitors. Service MNEs are therefore searching for competitive advantage, and this is usually a response to the requirements of existing clients or of potential customers. Competitive advantage can be achieved by the quality of the service supplied (a multifaceted attribute which depends on the kind of service and which is difficult to measure: see for example O'Farrell and Hitchens, 1990), the price of the service, and the support provided at the time of purchase of the service and for the duration of its utility to the purchaser (such as software updating, maintenance facilities, speed of updating, subscriber information services, response times to equipment failures etc.). In order to be able to exert some control on all or some of these variables, MNEs must establish ownership-specific advantages. Either they are able to obtain sole or privileged access to specific assets that allow a particular service to be provided at lower cost, or they should identify the best places from which to produce and deliver their services (locational choice advantages). Alternatively, they have to organize a set of complementary advantages more efficiently in order to generate lower prices or a better value-adding product (common governance advantages) (Dunning, 1989; see also Enderwick, 1989a). In these circumstances, agglomeration (clustering) of competing firms is an important locational consideration because the scope for introducing innovative services will be dependent upon the level of market development: the larger and more diverse it is, the more likely it is to be receptive to new kinds of services. Some of the consequences of these strategies for the relative development of nations and cities around the world will be considered in chapters 4 and 5.

Several competitive (or firm-specific) advantages can be identified for services MNEs. Many arise from the extent to which a service that provides specialist knowledge or information is able to provide it in state-of-the-art, accessible, easy to interpret form at the lowest possible cost. This is especially important for many of the information services such as securities dealing, stockbroking, data producing services or commodity broking. IT is a key part of both the production and the delivery of these services and therefore makes a vital contribution to their competitive advantage.

Economies of scale also induce firm-specific advantages: a large container ship with its own container handling equipment offers lower unit costs than several smaller, slower vessels without such equipment. The ability to retain highly specialized personnel will be determined by the scale of the organization or the extent to which they can be moved between different sections in order to allow economies of common governance. Risks can also be spread more widely if services operate as MNEs (for example some insurance services); funds can be raised at more competitive rates by large service MNEs (who can then perhaps price their services more competitively). Economies of scope arise when MNEs can provide a larger and more variable set of related services or can beat down the prices from suppliers because of the size of their operations. They are therefore partially a function of the spread of activities – hence the mergers by accounting and management consultancy firms in recent years (Leyshon, Daniels and Thrift, 1987).

Another source of competitive advantage for service MNEs is corporate identity. Certain brand names are associated with quality or product differentiation and, because many services involve the input of considerable human resources, there is a good deal of scope for variation. It is only necessary to think of legal services, retailing, car maintenance, or education to be aware of real or perceived differences in quality that determine our propensity to use them or to aspire to obtain a service such as education from one source rather than another. Many of the advantages or disadvantages attributed to using such services arise from our assessment of the personnel involved. This is, of course, crucial when the reputation of the client's service is likely to be affected by the quality of the supplier's services. Commercial firms are therefore very sensitive to the reputation, quality and prestige of the services they purchase – preferring, other things being equal, to use Young and Rubican (US), Saatchi and Saatchi (UK) or Dentsu (Japan) for their advertising portfolios and Arthur Anderson (US) or Peat Marwick Mitchell (UK) as their accountants or management consultants. In the case of consumer services MNEs, such as McDonalds (US), the Sheraton (US) and Holiday Inn (US) hotel chains, or Benetton

(Italy), the aim is to serve particular kinds of markets with a quality and consistency that can only be generated by their ability to marshal the best quality services (facilities) at lowest cost at multiple locations. Corporate identity has a part to play, and the fact that they are MNEs helps them to achieve these objectives and so retain or extend their competitive advantage.

Finally, variations in access to inputs or to markets can be crucial for the competitive advantage of service MNEs. It is clearly advantageous if a firm has knowledge of particular sources for inputs or knows how to identify such sources if and when required. Such an advantage is compounded if the MNE can also act as negotiator on behalf of the client at the location where the client wishes to use a particular service. This may therefore require direct representation of the MNE in the environments that are the sources of the knowledge and expertise that they assemble for clients. Very often there are complementary locations for a whole range of such services, giving rise to the agglomeration economies that, as we shall see later, are an important influence on the location behaviour of service MNEs. Advantages accruing from favoured access to markets can arise from early entry, often as a response to clients who themselves have been engaged in internationalization and who will only continue to use their regular service suppliers if they are as accessible as at the headquarters location. Others have anticipated existing and future client requirements: UK property consultants, for example, have established offices in several overseas markets so that they can advise clients on property investments, identify suitable properties for the direct use of clients and manage the overseas property portfolios of client companies. Some showed considerable foresight and began establishing overseas offices in the 1920s; others did so in the 1970s as the internationalization process gained momentum. Construction companies, international bulk shipping companies and design firms (table 2.5) have also expanded at international level as service MNE activity has increased. World fee income for commercial architecture and design was £66b in 1988, with 50 per cent of the market in the US. Commercial architecture accounted for £4.1b, interior design (including retail) £1b and product and graphic design (including corporate identity) £10.9b. Although international expansion of the largest companies accelerated in 1987 and 1988 they only command some £0.2b of the global income. The industry is still dominated by small, entrepreneurial companies.

Dunning (1989) distinguishes between favoured market access as a mechanism for responding to client needs or fulfilling the corporate strategy of the service supplier, and the setting up of overseas subsidiaries, import and export operations and buying agencies that essentially help to sustain the domestic activities of service MNEs. The competitive advantage stemming from this comes from their knowledge of markets, their bargaining power with suppliers, their ability to hedge against foreign exchange fluctuations and their ownership of trading and wholesale outlets. Retailing services are particularly appropriate for this form of firm-related advantage: good examples are Sears Roebuck (US) and the Japanese general trading companies.

Dunning (1989, table 2) provides a very detailed illustration of the ownership (competitive), location (configuration) and international-ization (coordinating) advantages of transnational activity for a wide range of service sector actors. Take the example of insurance services. Competitive advantages include image and reputation (Lloyd's of London) and specialized expertise in aviation or mari-time insurance, for example. Configuration advantages include the need to be in close touch with the insured parties and the economies of concentration for insurance suppliers (e.g. reinsurance); large insurers are striving to achieve oligopolies, government regulations in individual countries vary in the extent to which they allow direct imports of insurance services. Coordinating advantages arise from the scope for spreading large risks, or from government regulations that require local equity participation. In the case of advertising services there is a long list of competitive advantages including economies of coordination, image and creative ability, favoured access to markets (perhaps via subsidiaries of clients in other countries, including the home market), goodwill, and the value of making available the full range of services. The location (or configuration) advantages include cultural nuances in advertising style, presentation or expectation, the need to be close to mass media, and the necessity for face-to-face contact with clients if a good quality service is to be provided. The coordinating advantages for advertising services are also numerous: quality control, need for local imports, reduction of transaction costs by using agents rather

Table 2.5 Top ten international design firms, 1988

Firm	Fee income (£m)	Employees	Headquarters	Location of other offices
WPP Group (UK)	57.5	860	London	San Francisco, New York, Los Angeles, Tokyo, Manchester, Edinburgh
Lander Associates (US)	28.2	500	San Francisco	Los Angeles, Mexico City, New York, London, Paris, Oslo, Helsinki, Stockholm, Hamburg, Antwerp, Madrid, Rome, Istanbul, Athens, Seoul, Tokyo, Hong Kong, Bangkok, Singapore, Sydney
Michael Peters Group (UK)	22.5	720	London	Toronto, Madrid, Milan, Berlin, Helsinki, Tokyo
Addison Consultancy (UK)	20.0	300	London	San Francisco, New York, Singapore
Fitch-RS (UK)	17.4	500	London	San Diego, Columbus, Boston
Siegel & Gayle (UK)	16.5	230	New York	San Francisco, Los Angeles, London, Oslo, Copenhagen, Hamburg, The Hague, Madrid, Sydney
Pentagram (UK)	8.8	120	London	San Francisco, New York
Conran Design Group (UK)	8.5	240	London	Paris, Hong Kong
Wolff Olins (UK)	7.6	180	London	San Francisco, Copenhagen, Barcelona
Minale Tattersfield (UK)	5.0	55	London	Milan, Sydney

Source: Ramsthorn, 1989

Table 2.6   Classification system for service firm internationalization

| | Consumer/producer interaction: | |
	Lower	Higher
Relative involvement of goods:		
Pure services (low on goods)	1 Domestic mail delivery, show repairs	4 Engineering, advertising, education, insurance, medicine (III)
Services with some goods	2 Hotels, shipping, air frieght, retailing (II)	5 Personal air travel, maintenance (II, III)
Services embodied in goods	3 On-line news services, records and compact discs, computer software/disks (I, II)	6 Electronic mail, teleshopping (II)

[a] Clusters related to degree of investment, presence, control in host country: I = exportable; II = franchising, minority joint ventures, licensing, management agreements; III = foreign direct investment via branches, subsidiaries, mergers or acquisitions.
Source: derived from Vandermerwe and Chadwick, 1989, 82–4, figures 1 and 2

than establishments that are part of the same operation, globalization of products that utilize high levels of advertising (e.g. motor vehicles, washing powders, beers or photographic film), and national regulations. The advantages sought by legal services are in some respects similar to those for advertising, although they are often best delivered by the movement of lawyers to overseas clients or vice versa rather than by setting up overseas partnerships.

Multinational service companies therefore deliver services in a variety of ways and in a form ranging from pure services to embodiment in goods. This has led Vandermerwe and Chadwick (1989) to suggest a classification system for internationalizing services that reflects these variations (table 2.6). The resulting matrix produces a clustering of services, with each cluster having a different set of internationalization modes. There are six cells in the matrix, which may be summarized as follows:

1 low goods, ower interaction (such services may be provided anywhere and have limited international potential, e.g. mail delivery);
2 medium goods, lower interaction (goods that facilitate this

service are easily transported to foreign markets and internationalization potential is therefore high, e.g. retailing, air freight, hotels)

3 high goods, lower interaction (goods can be supplied internationally very easily, e.g. computer software on diskettes, video films, musical recordings on cassette tapes or compact disks)

4 low goods, higher interaction (labour-intensive internationalization where interface with customers is very important, e.g. engineering consulting, advertising, education services)

5 medium goods, higher interaction (internationalization requires balance between goods and people, e.g. maintenance systems, banking, personal travel)

6 high goods, higher interaction (service is provided on the basis of goods provision rather than people, e.g. teleshopping).

Three clusters emerge from this matrix. The first cluster is based on the export of a service through a good (cluster I: mainly services in cell 3). This requires the least investment, control and presence. The second cluster involves some dependence on third parties using franchising, licensing, minority joint ventures and related management arrangements (cluster II: some services in 2, 3, 5 and 6). A certain amount of investment is therefore required along with carefully planned control of the chosen method of involving a third party. The third cluster incorporates strategies based on FDI; interaction with clients is very important and internationalization therefore requires as much control as possible over the delivery of the service or the good embodying a service (cluster III: mainly cell 4 and some cell 5). Investment requirements are therefore very high, since the establishment of branches and subsidiaries or the pursuit of mergers and acquisitions is the favoured strategy for undertaking FDI.

Case Studies of the Development of Service MNEs

The geographical and organizational development of service multinationals has been traced in some detail in a number of recent studies (see for example Daniels, Thrift and Leyshon, 1989; Boyd-Barrett, 1989; Enderwick, 1989b; see also UNCTC, 1990). It is useful to consider some specific examples based on the case histories of three UK service multinationals in an attempt to illustrate the relationship between the theory (outlined above) and the practice as well as to underline the variability of the process according to the industry involved and its relationship with the regulatory environment. The range of possible case studies is large; the three chosen represent a spectrum from early involvement in internationalization (British Airways: air transport), through a very recent (1980s) international expansion (Clifford Chance: legal services), to the internationalization via diversification of a service conglomerate (Trafalgar House: oil and exploration, construction and production, civil engineering, commercial and residential building, hotels, passenger shipping and cargo).

British Airways

The airline industry is one of the most carefully regulated, with complex national legal and institutional arrangements designed to protect the domestic markets of national carriers. This held back the expansion of carriers within countries with geographical characteristics not conducive to growth of air travel (the UK, the Netherlands) or of the appropriate size but a very small population (Australia). There were therefore early and less restricted opportunities for providing intercountry and intercontinental services. Thus by 1988 the Australian airline, Quantas, was the sixteenth largest in the world according to revenue passenger kilometres flown. Almost all the flights are long-haul services outside Australia. The US deregulated its national airline industry in 1977 and this was followed by a rapid expansion in the number of carriers, intensive price competition and, more recently, a wave of mergers and acquisitions resulting in concentration into so-called 'mega-carriers' operating a 'hub-and-spoke' network system. It is expected that the European and Asia-Pacific airline markets will also be deregulated during the 1990s with similar results. National airlines (or flag carriers) have, at the same time, been privatized and must therefore find ways to survive and grow in increasingly competitive markets.

It is against this background that the internationalization of

British Airways must be considered. The origin of the company goes back to 1924 when Imperial Air Transport Ltd was established with the aid of government funds to counter foreign competition and to maintain air services to India and to Africa. During the 1930s these services were expanded to include the Far East (Singapore) and, via a partnership with Queensland and Northern Territories Air Services (Quantas), Australia. Meanwhile, in Europe two competitors to Imperial Air Transport were established in 1935: British Airways and British Continental Airlines. These were viewed as a threat and in 1939 the three companies were nationalized to form the British Overseas Airways Corporation (BOAC). In 1946 British European Airways (BEA) and British South American Airways (BSAA) were detached from BOAC; BSAA was remerged with BOAC in 1949. During the 1960s BOAC began to develop new air routes across the Atlantic to North America (with Cunard), to Japan and to the USSR. BEA diversified into holiday package tours (BEA Airtours Ltd). These initiatives returned both companies to profitability after a difficult period in the 1960s and they were merged in 1974 to form a new state-owned airline, British Airways (BA). In a step towards privatization the Secretary of State for Transport was the sole shareholder when BA became a public limited company (PLC) in 1984; the airline was fully privatized in 1986.

The newly privatized company obtained additional routes by taking over British Caledonian Airways (BCal) in 1987 and thereby extending its operating base from Heathrow Airport to the UK's second major international airport at Gatwick. The company is now organized into three divisions of subsidiary companies: airline operations, package holidays, and travel booking and finance. Some 23 million passengers and 361,000 tonnes of freight were transported between 163 city destinations in 1987–8 (table 2.7). The vast majority of these destinations are outside the UK, being almost evenly divided between developed and developing market economies. BA's international route network is one of the most extensive, covering most of Europe, the Far East, Australia and New Zealand, the Caribbean, Africa, India, the Middle East and North and South America. It was ranked seventh in the world by revenue passenger kilometres, freight tonne kilometres and revenues ($6.7b) in 1988.

Table 2.7   Geographical distribution of cities served and subsidiary companies, British Airways PLC, 1988

Type of economy and region	Cities served	Subsidiary companies
Developed market economies	100	23
North America	21	2
Western Europe	64	19
Pacific	9	2
Other developed countries	6	–
Developing market economies	58	1
Africa	22	–
West Asia	10	–
Asia (excl. W. Asia), Pacific, Oceania	13	–
Latin America, Caribbean	13	1
Centrally planned economies	5	–
Totals		
UK	15	18
foreign	148	6
overall	163	24

Source: British Airways PLC, Annual Report and Accounts, 1988

As one of the world's largest airlines, British Airways is constantly exploring ways of penetrating new markets or expanding its service network in order to ensure its role as a global airline: examples include its (abortive) attempts to form a partnership with United Airlines (USA) which would benefit from shared facilities and more coordinated links between internal and external schedules in the United States, or with Royal Dutch Airlines (KLM) which via its stake in North West Airlines would also have enabled more direct access to the US market. Most recently, BA has targeted US Air and seems unlikely to give up its globalization strategy. Efforts have also been made to take over or merge with major European airlines such as Scandinavian Air Services (SAS) or Sabena (the Belgian national airline). All these manoeuvres are an attempt on the one hand to consolidate BA's position in anticipation of deregulated European markets when it will need an airport with spare capacity (which Heathrow will not have), and on the other to gain access to a larger share of the US airline market by achieving control of a US carrier with a hub-and-spoke network. At present BA cannot provide its

Table 2.8 Divisional and geographical breakdown of turnover, British Airways PLC, 1984–1988

Breakdown of turnover	1984 £m	1984 %	1986 £m	1986 %	1988 £m	1988 %
Divisional						
Airline operations:	2,382	95.7	2,981	94.6	3,523	93.8
Scheduled service passengers	(1,906)	(75.8)	(2,376)	(75.4)	(2,858)	(76.1)
Scheduled service freight and mail	(208)	(8.3)	(268)	(8.5)	(287)	(7.6)
Non-scheduled services	(112)	(4.5)	(151)	(4.8)	(165)	(4.4)
Maintenance and services	(157)	(6.2)	(186)	(5.9)	(213)	(5.7)
Package holidays	79	3.1	120	3.8	217	5.8
Helicopter operations	43	1.7	38	1.2	–	–
Other	10	0.3	10	0.3	16	0.4
Geographical						
United Kingdom	382	15.2	423	13.4	378	10.0
Continental Europe	753	29.9	945	30.0	1,231	32.7
The Americas	670	26.6	990	31.4	1,175	31.2
Africa	154	6.1	169	5.4	237	6.3
Middle, Far East Australasia	555	22.1	622	19.8	735	19.6
Total turnover	2,514	100.0	3,149	100.0	3,756	100.0
Employment (no.)	37,247		40,254		43,969	

Source: British Airways PLC, Annual Report and Accounts, 1984, 1986, 1988

own services within the US to act as feeders to its US gateway airports such as New York, Los Angeles and Chicago. Needless to say, the US airlines are opposed to this strategy unless they can have reciprocal rights at European airports.

With some 50 per cent of its scheduled service passenger revenues coming from business travellers, British Airways has invested heavily in airport and flight services for this group and introduced new brands such as Club Europe and Club World. BA is also active in the development of computer reservation systems, such as the Galileo system which can then be sold or leased to other airlines. This activity is often undertaken jointly with other airlines: BA and four other airlines own 50 per cent of the Covia Apollo reservation system (used in the US) which will eventually be linked with the Galileo system. Likewise, BA also wholly owns Travicom which promotes and provides automated multiple access reservation services for the travel industry. A divisional and geographical breakdown of the turnover of British Airways and its subsidiaries is provided in table 2.8.

Clifford Chance

The international expansion of legal firms is much more recent than that of airlines. After the Second World War some US law firms did move to Western Europe; they were following foreign direct investment by US manufacturing firms such as Ford or General Motors. But the pace of internationalization in general was relatively limited until the 1980s and the globalization of securities markets. The complex financial transactions involving mergers, acquisitions, sales of equity and the issuing of corporate bonds required specialist legal advice and opened the way for European, North American, Japanese and Australasian corporate legal firms to open offices in other parts of the world. As with airlines, the precise form of the process has been governed by national regulations which vary with respect to the form of licencing to practise that is required, the legal services that can be provided and the form of the business relationship between domestic and foreign lawyers. Major banking centres have, not surprisingly, attracted a large share of the new offices of corporate legal firms serving overseas markets (see also chapter 4).

The Clifford Chance partnership (eighth in the world in 1988 by number of partners) was formed in 1987 by the full merger of two large London law firms, Clifford Turner and Coward Chance. The former was the larger and specialized in corporate finance and international law; it had a large network of associate relationships with overseas legal firms as well as its own international network of offices. For example, Clifford Turner specialized in providing advice on corporate transactions between Europe and North America and had already established a large office in New York in 1986. It had also operated in Japan for more than twelve years. In Europe the firm had offices in France (since the early 1950s), Amsterdam, Madrid and Brussels. Coward Chance diversified the expertise available in the new firm, into banking, finance and capital finance. It had also pursued a rather different strategy in its international expansion, being one of the first UK legal firms to open offices in the Middle East in the mid 1970s (in Saudi Arabia and the United Arab Emirates) and operating offices in Hong Kong (from 1980) and Singapore (from 1981). The rationale for the merger is therefore apparent: it provided an unrivalled geographical distribution of international offices combining the European operations of both firms, the US practice of Clifford Turner, and the Middle East and SE Asia networks of Coward Chance.

There is a network of offices in fourteen cities worldwide (table 2.9), and the 200 and 800 lawyers can provide legal expertise across a much wider spectrum of corporate requirements than each of the firms independently. The firm's particular strengths are in corporate legal work (including securities, mergers and acquisitions, venture capital, buyouts, company liquidation and reconstruction), international and domestic banking, legal matters (sterling issues, currency swaps, loan lending and guarantees, securities and remortgaging, property financing and off-balance-sheet financing), commercial litigation, legal services relating to EC matters such as customs laws, anti-trust policies, the General Agreement on Tariffs and Trade (GATT), anti-dumping law, and cases before the European Court of Justice. The Paris office is one of the largest corporate legal operations in France, while the New York office offers a European perspective for the North American market. These are just some examples of the diversity of the legal services that can be provided by a large multinational legal firm. The advice

Table 2.9   Number of offices and number of partners worldwide, Clifford Chance, 1989

Type of economy and region	Offices	Partners
Developed market economies		
North America	1	4
Western Europe	6	177
Pacific	1	3
Developing market economies		
West Asia	5	3
Asia (excl. W. Asia), Pacific, Oceania	2	8
Totals		
UK	2	153
Foreign	13	42
Total	15	195

Source: Clifford Chance, Brochure, 1989

available will be more comprehensive than that provided by smaller regional or local practices, and there will be scope to bring in the necessary expertise from other offices of the partnership if a specific need cannot be fulfilled locally.

Trafalgar House

The roots of the international expansion of Trafalgar House PLC are to be found in engineering and construction services. The market for construction services is prone to substantial fluctuations from high to low demand (see also chapter 5), and the international market moved from high demand during the 1970s (mainly for projects in the oil-rich Middle East states) to low demand during the first half of the 1980s, followed by an upturn after 1986, especially in Europe, North America and the Asia-Pacific region. In common with legal services, it is necessary for the factors of production to move to the sites where they are required; thus large-scale projects around the world that could not be fulfilled by domestic contractors were tendered for by overseas firms. In most cases they set up affiliates or establish joint ventures with local firms. Again, the precise form of the overseas operation will be governed by national rules on the

Table 2.10　Divisional and geographical breakdown of turnover, Trafalgar House PLC, 1984–1988

Breakdown of turnover	1984		1986		1988	
	£m	%	£m	%	£m	%
Divisional						
Property and investment:	178	11.0	352	17.0	676	25.3
Construction and engineering	1,045	64.8	1,299	62.7	1,362	50.9
Shipping and hotels	336	20.8	383	18.4	609	22.8
Oil and gas	53	3.3	36	1.7	28	1.0
Geographical						
United Kingdom	985	61.1	1,167	56.3	1,533	57.3
N. America and Caribbean	186	11.5	335	16.2	498	18.6
Africa	93	5.8	136	6.6		
Middle East	97	6.0	123	5.9	356	13.3
Europe	74	4.6	96	4.6		
Far East, Australia, India	179	11.1	215	10.4	289	10.8
Total turnover	1,613	100.0	2,071	100.0	2,676	100.0
Employment (no.)	31,249		33,278		27,814	

Source: Trafalgar House PLC, Report and Accounts, 1984–8

movement of professional, technical and construction personnel across borders, on regulations relating to the professional recognition and accreditation of the architects and engineers, and on access to sites etc. to make preliminary surveys prior to submitting bids for contracts.

When it was formed in 1956, Trafalgar House was only concerned with the residential and commercial property market. After going public in 1963, Trafalgar House embarked on organic growth via selected acquisitions that led to expansion into civil, offshore and structural engineering. As a result the group has won large contracts overseas as well as in the UK. It now has construction representation in the US, the Middle East and Latin America. The decision to diversify into shipping was made in 1971 when Cunard was acquired, including its flagship liner, the *Queen Elizabeth II*. Cargo operations were strengthened by the purchase of Ellerman Line in 1988. The company's portfolio was again widened in 1982 via oil and gas exploration and production in the North Sea and in the USA through the acquisition of Thomson-Monteith. The company develops office parks in the US, business parks in the UK, housing developments in the north-east US and major construction projects in Turkey, Egypt, Ghana, the US and Hong Kong (construction of the highly acclaimed Hongkong and Shanghai Bank building). The shipping, hotel and leisure interests extend to the West Indies and the US (hotels), to Australia, West Asia, the Persian Gulf, the Mediterranean and the US (cargo services) and to Scandinavia (cruise shipping). By 1988 Trafalgar House was the 35th largest construction company in the world by value of total contracts.

It seems reasonable to suggest that, apart from providing access to a wider range of international markets, diversification has enabled Trafalgar House to smooth out the effects of international and domestic downturns in demand for its construction services on its return from its investments. Although total employment had declined to 27,814 in 1988 (table 2.10) from 31,249 in 1984, the emphasis had begun to move away from the core businesses of construction and engineering services into property investment, shipping and hotels. Although primarily a UK-based service conglomerate, Trafalgar House has been expanding overseas during the 1980s, especially to the United States in its construction and engineering and its commercial and residential property divisions (table 2.11).

Retail Internationalization

While MNEs have by definition become heavily involved in internationalization, it is not universal across all the service sectors. Retail businesses, for example, have only cautiously embarked on internationalization even though firms like Sears Roebuck and Woolworth expanded outside the United States early in this century (Treadgold and Davies, 1988). The leading French retailing organization, Carrefour, first expanded into foreign markets in 1969, for example, with the opening of stores in the UK, Belgium, Switzerland and Italy, but the initiative failed. It did not expand again internationally until 1975 when it opened stores in Argentina, Brazil, Spain and the US. Since 1983 the number of domestic and foreign stores operated by the company has steadily increased but the ratio has remained at around three to one in favour of the former.

The success of retailers outside their national markets is heavily

Table 2.11 Number of offices worldwide, by division, Trafalgar House PLC, 1988

Type of economy and region	Commercial and residential property	Construction and engineering	Shipping and hotels	Oil and gas	Financial services
Developed market economies					
North America	5	13	–	1	–
Western Europe	23	46	26	5	8
Other	–	1	–	–	–
Developing market economies					
Africa	–	1	–	–	–
West Asia	–	2	–	–	–
Asia (excl. W. Asia), Pacific, Oceania	–	3	–	–	–
Latin America, Caribbean	–	–	3	–	–
Totals					
UK	23	40	24	5	8
Foreign	5	26	5	1	–
Total	28	66	29	6	8

Source: Trafalgar House, Shareholder Information, 1989

dependent on their ability to merchandise products with the image, price, quality and reliability, for example, which match the expectations of consumers in different cultural contexts. Highly specialized niche retailers serving exclusive markets, such as Burberry or Cartier, have for many years operated in the world's leading cities (New York, London, Paris and Tokyo, for example) but their share of the total retail market, national or global, is extremely small. Economies of scale and of scope are essential for the successful internationalization of merchandise retailers such as Marks and Spencer, W. H. Smith and C&A Modes.

Nevertheless the forces promoting the internationalization of retailing are increasing. National markets have in many cases become saturated and highly competitive; population growth in most of the developed market economies has slowed down to little more than replacement rate; and planning controls often restrict both the volume and the location of additional retail floor space and curtail the efforts of retailers to innovate in the marketing and provision of retail services. This means that opportunities overseas may, by comparison, look favourable, perhaps because the regulatory environment is less restrictive, the markets are relatively underdeveloped and there is a measurable need for the goods offered by the retailer. Most important, the returns from the overseas investment may ensure that the business continues to grow and offers a good return for investors: this may be especially the case if expansion involves the acquisition of retail interests in the target markets. This, however, is not a guarantee for success, as Marks and Spencer and other UK retailers who have tried to move into the North American retail market during the 1980s have found to their cost.

In general, it seems that the most successful attempts at developing an international presence are undertaken by niche retailers. Several examples are cited by Treadgold and Davies (1988) including Athlete's Foot, Body Shop, Levi, Sock Shop and Benetton. The last more than doubled its outlets worldwide between 1982 and 1986 (table 2.12) after opening its first shop in Italy in 1968. Treadgold and Davies (1988) suggest that a strong trading format, a clearly differentiated product (in quality as well as design) and a firm belief in the universality of the product explain the success of Benetton and similar retailers for whom the international

Table 2.12   International expansion of Benetton, 1982–1987

Country	1982	1986	Change
Italy	659	753	94
West Germany	138	321	183
France	198	376	178
Great Britain	35	239	204
Netherlands	8	41	33
Belgium	12	22	10
Austria	28	52	24
Switzerland	53	84	31
Other Europe	26	312	286
USA	0	523	523
Canada	0	41	41
Lebanon	0	4	4
Israel	0	4	4
Other Middle East	0	10	10
Australia	0	49	49
Hong Kong, Singapore	0	8	8
South Africa	0	8	8
Caribbean	0	3	3
Other	79	0	−79

Source: after Barry and Warfield, 1988, given in Treadgold and Davies, 1988

arena has become the domestic market. Excellent examples in food retailing are McDonalds, Burger King and Kentucky Fried Chicken, whose services and products are so precisely specified that they seem to override the kinds of differences in client taste, fashion and expectations that have caused problems for mass merchandise retailers as well as some of those attempting to internationalize on the basis of niche strategies.

Government Influences on the Tradability of Services

A very important influence on the global development of IT networks, on transport investments and on the strategies of service firms is the role of government (federal, state, provincial or local). Self-regulation by professional associations or institutions such as Lloyd's on the London Stock Exchange is also often recognized and monitored by governments at the appropriate level. Manufactured goods and primary products have long been the subject of regulation, incentive or control by governments with demonstrable effects on trade patterns and the flows of FDI. The oil crisis of 1973–4 triggered a shift towards greater trade liberalization as countries tried to generate revenues that would compensate, at least in part, for the crippling increases in oil prices. But as the destabilizing effects of the crisis diminished during the early 1980s there was a return to greater protectionism using non-tariff rather than tariff barriers to trade. According to Riddle (1986) the increase in non-tariff barriers affecting services has been geometric; the 2000 barriers identified by Anderson (1982, quoted in Riddle, 1986, 194–5) were 'designed to protect from foreign competition the services provided by local banks, communication monopolies, insurance companies, transportation companies, data processing organizations and other service industries. Many of these discriminatory practices strike at the heart of international businesses by impeding or even blocking the flow of vital management information across national boundaries'.

Services have an important strategic role in national economies (see for example UNCTAD, 1989a; Riddle, 1986). They help to sustain national sovereignty and security through a wide range of services, armed and otherwise, provided by government employees. Services are the source of the basic infrastructure for the daily requirements of the population and the functioning of the economy: highways/railways for commuting and freight movements, telephone services, emergency services and the police. Government-funded education and training programmes help to maintain the human capital that is the key resource for many service industries, especially information-intensive activities. Both of these aspects will have implications for the location of decision making functions (and of production) within the national space economy. This, in turn, will contribute to the competitiveness of national firms in global markets.

Traditionally, governments not only have been assumed to carry responsibility for services of this kind, but also have been the principal, if not the sole, supplier. Government ownership (and monopoly) was considered the only way to ensure that strategic services were available where and when they were needed; the government exchequer was also the only purse large enough to find the massive investment in infrastructure. As national economies have become more interdependent, service and manufacturing companies have been exposed to more intense and searching competition than ever before. Strategic services are incorporated in the production function for firms, and any inefficiencies in the quality, cost and availability of these services will ultimately be reflected in the competitiveness of national firms. As a result, many governments actively promote the creation of a more competitive environment for the provision of strategic services, mainly through the privatization and restructuring of government-controlled activities (such as water supply, telecommunications, or air transport in Britain) to create a more competitive supply. At the same time governments have become more involved in the regulation and/or deregulation of services with a view to increasing availability on a more equitable basis or in a way that proves attractive to the location of foreign service firms. A recent example in the US is that the Treasury is considering a major overhaul of its banking system (Durie, 1990). This would allow outside (non-banking) corporations to own banks and would dismantle barriers within the financial services sector. Many of the US restrictions have their origins in the Great Depression when it was considered desirable to divide banking functions from those ascribed to the securities industry so that the bank would be protected if the securities arm collapsed. Japan has a similar division and is under increasing international pressure, along with the US, to make changes that move towards the universal banks that have developed in the UK following the Stock Exchange reforms of 1986 or have existed for much longer in Germany. Other US restrictions on banks owning branches across state borders may also be ended (some states have already set up reciprocal arrangements with other states allowing interstate banking), and overseas banks, which find it easy enough to get into the US, will be able to compete on a more comprehensive basis with domestic banks.

It seems, however, that non-tariff barriers are being increasingly used to inhibit the movement of tradable services. For developing countries, non-tariff barriers are a way of protecting domestic service industries (and therefore job availability for nationals) from direct competition with more cost-effective and innovative foreign suppliers of similar services and discouraging the direct involvement of superior overseas suppliers (largely from developed market economy countries) in national markets. As a result, foreign service MNEs are often treated less favourably than national firms. But non-tariff barriers may be used selectively. Some countries are anxious to attract inward investment in services, especially for large infrastructure projects for which national expertise (as well as resources) is not available. Brazil has sought foreign MNEs to participate in massive port and highway projects, Greece has sought foreign bids for investment in a new airport for Athens, and Singapore has created a regulatory environment for financial services that involves minimal interference in their activities. The emergence of positive attempts to encourage certain 'target' or 'strategic' services in line with particular national needs is a recognition of the danger of over-protecting national strategic services: that is, their reduced exposure to innovations in production and delivery will ultimately affect the competitiveness of the national economy as a whole as well as specific sectors.

There are four major types of non-tariff barriers to trade and investment in services: on product, capital, human movements and establishment (Riddle, 1986; see also Krommenacker, 1984) (table 2.13). Movement of a service may be confronted by total exclusion from a national market, by exclusion unless a certain proportion of the service product is either produced or wholly reproduced in the country, and by limitations on connections between private international telecommunications networks and public networks. Some countries restrict foreign advertising or require that any advertising is initiated and produced in the domestic market (e.g. Canada). Limitations on where government-funded activities such as defence services or educational establishments can purchase their equipment or technology will restrict product movement. Complicated procedures for receiving and distributing products or obtaining verifications that meet national standards (a frequent problem for importers to Japan) are ways of discouraging unwanted product movement. Particularly significant for the tradability of services are barriers to human movements, to the movement of capital and to the rights of foreign service firms to establish a presence in a national market. Human movements are discouraged not only by the use of work permits or visas but by non-

Table 2.13 Types of non-tariff barriers to trade in services

| | Trade services | | Investment services | |
Barrier to:	Across border	Domestic establishments	Foreign earnings	Third country
Product movement	Market access Local purchase Telematics Govt activity[b] Technical standards Charges/taxes Intellectual property	Telematics[a]	Market access Local purchase Telematics Govt activity Technical standards Charges/taxes Intellectual property	Market access Local purchase Telematics Govt activity Technical standards Charges/taxes Intellectual property
Capital movement	Currency restrictions	Currency restrictions	Currency restrictions Repatriation of profits	Currency restrictions Repatriation of profits
Human movement: labour, consumers	Work permits	Visas Departure tax	Work permits	Work permits Visas Departure taxes
Producer establishment			Right of establishment Access to production inputs	Right of establishment Access to production inputs

a In consumer's home country.
b Subsidies, dumping, procurement practices, regulations, monopolies.
Source: Riddle, 1986, 200, table 9.2

recognition of professional qualifications gained in one country as the basis for practising in another (e.g. in the medical and legal professions). Consumers may also be discouraged from crossing borders to benefit from lower prices (thus undermining demand within their domestic economy) or duty-free goods by levying high departure taxes (e.g. Indonesia) or protracted border customs controls. Capital movements have always been the engine of international trade and have become more so since some financial and professional services have been transformed into MNEs. But exchange rates and interest rates influence the flow of capital, along with limits on the expatriation of foreign earnings. Information technology has, on the other hand, made it much more difficult for governments to implement barriers against global currency trading or financial transactions either within or between organizations (using public or private telecommunications channels). Rights of establishment by foreign service firms in national markets can to some extent be achieved by putting up barriers to human movement. If these are not adequate then ownership requirements can be restricted, e.g. until 1987 the maximum foreign ownership of members of the London Stock Exchange was 29.7 per cent. The establishment of subsidiaries, especially of banks, may be totally prohibited or restricted to a specified quota (e.g. Canada). Even if it is permissible for firms to establish offices in another country, they may still be required to employ a specified proportion of local staff or local employees with specified skills.

The tradability of services is also heavily influenced by government regulations intended to enforce standards (Gibbs and Hayashi, 1989). Such standards may pertain to the compatibility of civil aviation facilities and maintenance arrangements or of electronic data exchange technology; to the content and cultural affinity of advertising; to the practice of professions such as accounting; or to the operating systems for personal computers. Without international harmonization, world markets for services will be fragmented and relatively inefficient. The prospects for the greater involvement of developing countries wishing to expand their service exports or to introduce more liberal regulations conducive to service trade (Gibbs and Hayashi, 1989) are also affected by the activities of specialized international organizations concerned with standards. The technical standards set by the International Telegraph and

Telephone Consultative Committee (CCITT) and the International Telecommunications Union (ITU) are intended to encourage worldwide conformity, and the Electronic Data Exchange for Administration, Commerce and Transport (EDEFACT) has promoted international harmonization in the telecommunications and transport field. The Chicago Convention on Civil Aviation has sought to standardize a wide range of international aviation practices and policies. Attempts at achieving harmonization of standards are more straightforward in some areas than others: standards for professional practice and qualifications are complicated, for example, by linguistic, cultural or religious considerations. The pace of technological change also poses problems in that it tends to be geographically uneven, so undermining international standards and widening the differential between nations, intensifying the barriers to market entry and helping to sustain captive markets.

As individual national governments work to create the conditions that will protect or promote tradable services (or a combination of both), some are establishing bilateral and multilateral trade agreements. One example was the introduction in 1992 of a unified market for capital, persons, establishments and services (as well as goods) within the European Community as intended by the Treaty of Rome (1957). The measures introduced number 285 in all and are laid down in the Single European Act (1986) which has been adopted by the twelve member countries of the Community. The markets for services in Europe have been highly fragmented and certainly highly protected by a variety of national non-tariff barriers. The new measures are intended to require the development of conditions that will allow service firms (and the individuals who provide services) of any member state to sell those services in any other (or all) member states without restriction. This will require professional qualifications, such as in law or architecture, obtained in one country to be acceptable in any other member state, so that truly common markets in banking, financial, professional and certain transport services (especially air transport) can then exist.

The Cecchini (1988) report lists five effects (at the macro level) of the Single European Market. First, the removal of cross-border controls will reduce delays and administrative bureaucracy on the movement of services (transport is the most obvious beneficiary but movement of labour will also be simplified). Second, the cost of purchasing services by public enterprises or public administration will be reduced following liberalization of procurement markets. Third, a larger market will allow service businesses to allocate resources (or to rearrange production) in a more efficient way and so bring about price reductions (following the effect of economies of scale on unit costs, for example). Finally, liberalization of financial services should reduce the prices charged although this should be compensated by increases in volume of business.

Thus, in the case of banking, the Second Banking Directive (1989) adopts the principles of harmonization, national recognition and home country control (Evans, 1990). The intention is that this will result in a more competitive European banking market: banks of any country will be able to market and provide their services anywhere else in the Community. An idea of the scale of the European market is provided by the fact that European Community banks have assets twice the size of US banks (Evans, 1990) and one-third of the assets held by the world's top 500 banks. Thus the mutual recognition of a banking licence granted in one country provides a major opportunity for banks to provide lending, financial leasing, deposit taking and other forms of borrowing, trading in foreign exchange or financial futures (for example), money broking, portfolio management and advice, participation in the issue of shares and other banking services without restriction in any of the other eleven member states of the Community. The interesting geographical problem is how best to trade these services: the possibilities are to acquire other banks in the larger markets (countries), to directly establish a network of branch offices (organic growth), to arrange reciprocal capital arrangements with other banks, or to specialize in certain services in order to develop 'niche' markets. Some strategies will encourage greater dispersal of banking activities in Europe; others, such as reciprocal capital arrangements, will probably maintain established patterns of banking activity in a small number of major European cities. Acquisition of other banks may lead to further concentration of banking control in certain locations within Europe. Some member states have banking systems that are better placed to take advantage of the Second Banking Directive, notably the UK, while other countries such as Spain, France and the Netherlands have found it necessary to increasingly liberalize, rationalize and (where appropriate) privatize their banking system as the need to operate in a more competitive banking environment approaches (for a detailed discussion see Leyshon and Thrift, 1992).

Some bank services can be marketed and supplied from a home country base without direct representation in the targeted national market. This is not true for insurance: it is more likely to rely on physical establishment either directly by the purchase of a local insurance company, or via an agency arrangement, or through a representative office. Many kinds of insurance require direct customer contact either at the establishment or by the use of representatives travelling to clients on a regular basis. Clients may adopt a rather negative view of a foreign insurer that does not have a local office (Diacon, 1990). In common with banking and similar financial services, the Single European Market will allow individual insurance companies to maximize any competitive advantage they have from using information technology.

The internal market is intended to reduce income differentials between member states, to harmonize the kind of products required, and to induce greater similarity in business practice and in marketing. All these things will generally promote the tradability of these and other services within the European Community. It remains difficult, however, to know what this will mean in terms of cost savings for firms using different types of services. One study of the cost to business services of not moving to a Single European Market (Commission of the European Communities, 1988) estimates the higher cost of business services (the direct effect) at up to 3.5b European Currency Units (ECUs), with three indirect effects – lower output in the whole EC economy, lower demands for business services, and higher costs from forgone externalization of the supply of business services by other firms – amounting to 0–3.0b, 0.8–2.5b and 0.1–0.2b ECUs respectively. The programme is also very ambitious concerning the time available to agree such a large number of measures. Some, such as the harmonization of value-added tax (VAT) throughout the Community, are very contentious and may take many years to achieve (Treadgold and Davies, 1988).

A second example of trade cooperation between countries is the Canada–US Free Trade Agreement of 1988, which was signed after negotiations extending over six years. It comprises written agreements concerning trade in services between the two countries and the rights accruing to each with respect to maintaining foreign establishment in support of service trade (Stern, Trezise and Whalley, 1987; Harrington, 1989). The Agreement (which includes other activities as well as services trade) covers services in general, investment, financial services and temporary movement of business personnel between the two countries. Harrington (1989) lists the service provision that companies from the home country can expect in the other country (the host country):

1 treatment according to the host country's general rules for the conduct of that service (this is often described as national treatment)
2 the right to establish support for the provision of their service in the host country (a distribution network, direct branch representation or agency)
3 licences or certificates using criteria identical to those for companies or professionals indigenous to the host country
4 unrestricted entry of home country managerial and professional personnel required to support the supply of the service in the host country
5 free entry (i.e. no tariffs levied) of services produced in the home country for sale in the host country
6 absence of any restriction on investment in services in the host country (although there is a threshold for US investment in Canada that is to be reviewed if it is reached within three years)
7 no limits on profit repatriation.

Although this Agreement represents a considerable advance on previous service trade arrangements, a great deal still remains to be achieved with respect to the liberalization of existing barriers (Harrington, 1989). Unless specifically incorporated in the Agreement, any entry barriers in existence prior to its ratification that were discriminatory to services from outside the host country will continue. Thus, limitations on the foreign ownership of the small number of very large Canadian banks will remain, and workers in clerical and secretarial occupations (these are not specifically listed in the Agreement) will continue to be prevented from moving across the border on their companies' business. A further difficulty that will require future attention is that the Agreement has been negotiated by two federal governments but some of the services involved are actually controlled or licensed by provincial or state governments.

Changes in Consumer Requirements and Expectations

As a wider range of service industries have become increasingly international in their operations and as technology has improved access to clients, it has become even more important to properly assess consumer needs and expectations. Until the mid 1980s the value of customer service was not seriously addressed by most service suppliers; their main concern was how best to handle the managerial problems presented by the development and delivery of services (Blois, 1983). Relatively little was known about consumers' perceptions of services, or about their assessment of the 'value' of services and of the effectiveness of services in meeting their particular needs. But at least two factors have made customer service a competitive weapon (Christopher, 1984). First, customers' expectations have been rising steadily as they have become more informed about the abilities and the limitations of any particular service for the purpose in hand. A good example is the widespread adoption of computers for data processing by many large companies from the mid 1960s onwards. The main concern was to acquire the hardware (often accompanied by some internal corporate reorganization); later came the realization that this hardware was only really as good as the software used to provide the processing service. For example, customers realize that the real value of computers is in the speed, sophistication, power, flexibility etc. of the software, and this is not a quality provided in equal measure by all software producers. The second factor that has raised the profile of customer service is the growing imbalance of added value as many service products converge because they are essentially based upon the same technology. Most purchasers find it difficult to perceive much difference between many personal computers produced by different manufacturers, which can be operated using the same software. What then becomes important is the ease with which a particular model can be obtained, the ease with which 'add-ons' can be plugged in, the compatibility with remote mainframe computers connected by telephone lines, the extent to which a computer can interface with a fax machine, and the speed of response by the supplier to system breakdown or to problems associated with loading and running complex software.

One of the effects of the enhanced demand for product support is to encourage services to seek economies of scale by serving more extensive but more specialized markets. This fits with the evidence for greater tradability of services and internationalization. Whether firms can succeed will depend on their success in identifying relevant customer needs. These needs can be represented as a hierarchy which exists with some consistency across cultures and is therefore relevant to the further development of global markets for services (Riddle, 1986). Customers require comfort (business surroundings, readily accessible location, easy parking); security (the service is guaranteed to be available, fair treatment); belonging (group membership); status (recognition as a client, member of an elite group); authority (the service is provided with particular attention to the needs of each individual consumer, and can be consumed according to the requirements and objectives of the user); and self-actualization (knowledge enhancement, altruistic results). The way in which each of the implicit service components represented by each of these needs is actually delivered does of course depend on how each customer perceives them; this will vary in relation to personality and to culture. A service considered perfectly accessible by customers in one part of Europe may be considered less so in some other part of Europe where perception of distance is different. In one culture security may follow from taking care of a service for oneself; in another it may represent a hands-off approach in which the supplier takes full responsibility for all aspects of delivery.

In an effort to retain a competitive edge in more 'customer aware' markets which are also likely to be more geographically and/or culturally diverse, the explicit service components can be modified by adjusting availability, (e.g. by extending hours so that information services are available 24 hours a day, 7 days per week, or updating such information every 12 hours; providing more locations, e.g. through automatic teller machines (ATMs), or by bringing car servicing to the home rather than expecting the car owner to take the vehicle to the repair shop; and extending access to information such as the teletext services available in the UK, France, Japan and the United States. Service producers can make changes to the environment in which they offer their service by providing specialized facilities, or by franchising based on standardized environments of the kind provided by motel chains in the US and Canada or by fast-food restaurants around the world. A third way of attaining a competitive edge, often dictated by the need to overcome cultural obstacles in global markets, is to set up self-service arrangements: for ticketing on rapid transit or metro services, for buying in retail supermarkets, for identifying faults with electrical equipment (using diagnostic guides which may be embodied in software in the case of personal computers), for diagnosis of personal well-being or health using diagnostic kits (including those for pregnancy), and for step-by-step guidance to buying and selling a house without recourse to a legal adviser or to importing a car from some other (cheaper) overseas source.

While all this suggests that flexible specialization is the key to competitiveness in service delivery (see for example Schoenberger, 1988; Gertler, 1988), it has been argued that a production line (or Fordist) approach can also reduce the final cost of delivery per unit (Levitt, 1976). Finally, competitiveness can be achieved through modifications to the range of a service. This may take the form of additions or modifications that make it possible to utilize the service in a wider range of circumstances, e.g. the development of laptop computers allows business travellers to word process on the move and to produce output that is compatible with desktop machines back at the office. Customization and/or specialization also change the scope of services while offering services in places where they were not previously available: a notable example of this is the appearance of financial services or vehicle maintenance services within hypermarkets formerly devoted only to the provision of retail goods.

One of the dilemmas for service firms with international ambitions is how far to respond to cultural differences in customer needs and expectations. One approach would be to adjust the delivery to meet expectations in different countries but to standardize as much as possible. This will obviously ensure that benefits from economies of scale are protected. But as the relative costs of international travel have fallen and average disposable incomes have increased, a larger proportion of the population have international experience; this may include the use of a 'home' service in another country. Similar standards of quality, quantity, environment and style of delivery are expected irrespective of the location of the 'host' supplier.

THREE

International Trade and Foreign Direct Investment in Services

Introduction

As world trade has grown steadily over the last 40 years there has been a significant deepening of global economic interdependence. Although the circumstances are complex, the greater tradability of services in general and producer services in particular during the last fifteen to twenty years has contributed to this trend. For the reasons outlined in the previous chapter it is reasonable to expect the link between increasing tradability and globalization of services to continue, especially if widespread liberalization of constraints on capital flows and on labour continues and the ideological belief in the efficiency of the market for allocating scarce resources in the industrialized countries remains pre-eminent.

Two of the principal measures of this process are international trade in services and foreign direct investment (FDI) by services. This chapter will examine some of the determinants of the geography of this trade, will explore some of the contrasts in the composition of services trade and FDI for developed, less developed and socialist economies, and will consider some of the issues involved in the liberalization of trade in services as part of the ongoing negotiations during the protracted Uruguay Round of the General Agreement on Tariffs and Trade (GATT).

Before proceeding, it is worth summarizing some of the characteristics that distinguish trade in services from trade in goods. First, the processes of production and consumption of many services must occur simultaneously and at the same locations; direct contact between producer and consumer is also required in some cases. Second, we have seen how the operations of many service industry markets are regulated, both directly and indirectly, by national authorities. Third, the obstacles or barriers to trade in services are mainly non-tariff (i.e. do not involve a clearly specified add-on charge of the kind levied on the importing of a merchandise product to country A from country B). It is much more difficult to quantify the cost disadvantages to services trade between country A and country B as a result of government regulations directed at, for example, the operations of foreign insurance, banking, information services, advertising or broadcasting services.

Measuring Trade in Services

It remains difficult to compile an accurate picture of world trade in services, but it is possible to provide an indication of broad orders of magnitude (UNCTAD, 1989a; GATT, 1989). In 1986 world receipts by major balance of payments components amounted to more than $3 trillion, of which more than $1 trillion comprised receipts from property revenues, interest income, labour revenues, official transactions, travel and other services. The International Monetary Fund (IMF) compiles balance of payments statistics for member countries (expenditures on imports, receipts from exports) which can be used to compile a cross-section of the value (by country) of service exports and service imports in 1984. The IMF disaggregates these data into shipping, travel, passenger services, other transportation and other services. Such disaggregation is clearly far from satisfactory since 'other services' includes everything except transportation and therefore many of the fast-growing producer services which are of particular interest here are not separately identified.

The volume of international trade in services is almost certainly underestimated in current balance of payments statistics (GATT, 1989). Intrafirm trade in services by MNEs, for example, is not adequately captured in official data (Markusen, 1989; see also DeAnne, 1990). MNEs are heavily engaged in exporting a variety of producer services from home (or headquarters) locations to subsidiaries, but these transactions are not identified as such because they are not between legally separate firms. A large proportion of the financial payments from subsidiaries to parents are for engineering, marketing, financial, computing or management services (see for example Caves, 1982; Mansfield and Romeo, 1980). The likelihood that such internal trade is significant makes economic sense in that the marginal cost of transferring knowledge-based assets between geographically dispersed locations is low. Since many MNEs utilize knowledge-based assets they will be engaged in internal trade of those assets. Markusen (1989) takes the view that while balance of payments data will include some intrafirm trade (if it is invoiced as such e.g. fees and royalties), most is subsumed under returns to FDI which are classified separately from trade in producer services. The result is that international trade in producer services is under-estimated because of the practical difficulties of identifying the part of FDI attributable to retained earnings (returns on knowledge-based or financial capital). It would be worth attempting to measure this in order to arrive at better estimates of trade in producer services.

Trade is also measured in different ways depending on the service activity involved (Gibbs and Hayashi, 1989; Segebarth, 1990). Accounting, corporate legal services and advertising measure their transactions on the basis of billing or fees, while construction services use the value of contracts. Some of the fees (or billings) will represent trade but others will not since they will be made to local market clients. Similarly, in financial services it is very difficult to disentangle flows of capital, such as the massive foreign exchange transactions that occur between the world's major financial centres every day, from payments for financial services provided to clients outside the home country. There are also large international transactions that involve insurance services but the trade element is again difficult to isolate; perhaps it should only include payments for transport insurance, e.g. to Lloyd's of London by foreign companies chartering or owning oil tankers (or by international airlines for their aircraft fleets). Not only is information on international trade in services insufficiently disaggregated, it is also virtually impossible to map the origins and destinations of trade flows. It is also difficult to tell from the available statistics where exactly national advantages reside for trade in services. Why, for example, should France be showing a growing share of services in its exports in the late 1980s, while for the UK the proportion has declined (see below)?

The industry classification used by the IMF does not adequately distinguish different types of services trade. Three basic types have been suggested by Grubel and Walker (1989) in their study of international trade in services between Canada and the United States (table 3.1): trade in factor services, trade arising from the temporary movement of people and goods, and trade in embodied services. Factor service trade is the best known form of international trade: it comprises the earnings from assets held outside the home country. Such earnings come from short-, medium- or long-term use of assets or in the form of debt and equity which generate interest, dividends or reinvested earnings. It should be apparent that factor service trade is closely linked to the flows of capital at the international level, and these are recorded in the balance of payments statistics of the kind gathered by the IMF. Table 3.1 shows that earnings from overseas assets represent some 4 per cent of Canadian GDP, a much higher proportion than for the US (less than 1 per cent). This is a symptom of the asymmetry in services trade between two countries with populations and economies that are greatly different in size and complexity. Returns from intellectual capital in the form of fees and royalties on books, recordings, films, computer software or medicines, for example, are another component of trade in factor services. These are a much smaller proportion of both US and Canadian GDP although, as with returns from foreign assets, they are rather more significant for the Canadian economy.

The delivery of many services incorporates persons who carry the knowledge or the expertise that comprises the service (accountants, engineers, software specialists or surgeons). Trade arising from temporary movement of people and goods requires crossing a border either to make use of (or to absorb) a service (a tourist or a postgraduate student) or to deliver a service. Movements of less than six months' duration are included in this category; longer-term movements are classed as migration flows. Goods in the form of capital equipment owned by companies in the US and placed temporarily in Canada to provide a service can also be included in the category. There is a balance in favour of the US in this trade, largely because more Canadians enter the US as tourists than vice versa. The second form of trade involving movement of people is

Table 3.1 Service trade between Canada and the United States, 1983

			Imports		Exports	
Type of service trade	Imports[a] ($m Can.)	Exports[b] ($m Can.)	Canada (% Can. GDP)	US (% US GDP)	Canada (% Can. GDP)	US (% US GDP)
Trade in factor services:						
Return to foreign assets	14,302	2,518	3.52	0.38	0.62	0.07
Fees and royalties	1,281	453	0.32	0.03	0.11	0.01
Trade arising from temporary movement of people and goods:						
Travel and transportation	4,934	3,477	1.22	0.13	0.86	0.09
Other private services	792	433	0.20	0.02	0.11	0.01
Trade in embodied services	18,012	20,058	4.43	0.48	4.94	0.54

[a] Value of imports from US to Canada.
[b] Value of exports from Canada to US.
Source: derived from Grubel and Walker, 1989, 207, table 7

generated by business services, the largest component of 'other private services' in table 3.1. Although a relatively small part of services trade, it is important to both the US and Canada (and the countries involved with business services) because of the opportunities it offers for the future and because it is partially linked to trade in goods.

It is also possible to define trade in such services in so far as they are embodied in shipments of material substances (Grubel and Walker, 1989). Thus an architectural consultant can trade his service to a host country after he has completed the drawings and specifications and transmitted them through the post, or electronically by facsimile or computer link, to the client. Similarly, a personal computer shipped from Japan to the United States embodies a range of producer services in its design, manufacture, operation (software) and maintenance. In other words, a certain proportion of the value added to every good or material is created by producer services. It is very arbitrary to treat a disk holding a computer program as a goods export and a disk holding an economic report as a service export. But how can embodied services trade be measured? We need to know what fraction of a unit of final output of a good is attributable to a service. Momigliano and Siniscalco (1982) have used input–output tables to try to estimate the indirect and direct service inputs per dollar of final output for different industries. These estimates can then be applied to each industry's exports and imports. Grubel and Walker (1989) assume that a dollar's worth of Canadian and US traded goods contains 30 cents of embodied services. The result is that embodied services trade is very significant (table 3.1) – equivalent to 9.4 per cent of Canadian GDP.

An Outline of Global Trade in Services

Almost 81 per cent of the total exports of services in the world economy originate from just twenty countries (table 3.2). The same countries, although in a rather different rank order, are amongst the top twenty importers of services (77 per cent of total world imports). The United States dominates global service transactions: more than one in ten transactions (by value) either originate there or are destined for US customers. Only France (for service exports) and

Table 3.2   Top twenty exporters and importers of services[a] in the world economy, by country, 1984

Country	Exports Value ($b)	Share of world total (%)	Rank	Imports Value ($b)	Share of world total (%)	Rank	Balance of exports minus imports ($b)
USA	45.6	12.1	1	46.2	11.5	1	−0.6
France	39.2	10.4	2	29.1	7.3	4	10.1
UK	29.5	7.9	3	25.6	6.4	5	3.9
Germany, FR	26.0	6.9	4	37.0	9.2	3	−11.0
Japan	24.6	6.5	5	42.3	10.6	2	−17.7
Italy	19.0	5.1	6	14.4	3.6	7	4.6
Netherlands	18.1	4.8	7	14.7	3.7	6	3.6
Spain	14.0	3.7	8	5.7	1.4	14	8.3
Belgium/Lux.	11.9	3.2	9	10.5	2.6	11	1.4
Singapore	9.7	2.6	10	−[b]	−	−	−
Austria	9.1	2.4	11	5.0	1.2	18	4.1
Canada	8.6	2.3	12	12.2	3.0	9	−3.6
Sweden	7.5	2.0	13	7.3	1.8	13	0.2
Switzerland	7.3	1.9	14	4.7	1.2	20	2.6
Norway	7.2	1.9	15	10.5	2.6	10	−3.3
R. of Korea	6.4	1.7	16	5.3	1.3	17	1.1
Denmark	5.9	1.6	17	5.7	1.4	15	0.2
Mexico	5.8	1.5	18	5.4	1.3	16	0.4
Australia	4.9	1.3	19	7.3	1.8	12	−2.4
Saudi Arabia	4.4	1.1	20	14.2	3.5	8	−9.8
Top 20 countries	304.9	80.9	−	303.1	75.4	−	−3.1
World total	376.0	100.0	−	400.4	100.0	−	−24.4

[a] Shipping, travel, passenger services, other transportation, 'other services' (as used in IMF balance of payment statistics).
[b] Singapore not included in top twenty importing countries.
Source: compiled from UNCTAD, 1989[a], 275, annex table 4, and 277, annex table 5

Japan (for service imports) can exert the influence of the United States. As a result, France was the leading net exporter of services by value in 1984 while Japan had the largest net deficit of imports over exports of almost $18b. In common with merchandise trade, there are some marked imbalances in services trade: it is dominated by developed market economies (only four developing countries are listed amongst the top twenty importers/exporters in table 3.2), with

44 per cent of the exports and 45 per cent of the imports generated by just five countries (the United States, France, the United Kingdom, Germany (Federal Republic) and Japan).

More recent estimates by GATT (1989) show that exports of services have continued to grow: the total value was $560b in 1988 and it will probably have exceeded $600b in 1989. Italy had displaced Japan amongst the top five exporters by 1988 but the list of leading importers is effectively unchanged. The EC is well ahead of the US in the international export of services: almost 22 per cent of total international EC trade is accounted for by services, whereas the equivalent statistic for the US is only 16 per cent (Segebarth, 1990). The member states of the EC operate a surplus (almost $13b in 1985) from services, while Japan (whose share of total world trade in services is increasing steadily) continues to have a deficit ($14b in 1985). The balance of payments for US service trade fluctuates between deficit and surplus (Segebarth, 1990). There appears to be a relationship between the level of economic development of a country and the propensity to export services or to generate demand for service imports. In view of the explanations advanced in chapter 1 for the expansion of services during the 1970s and 1980s it should not be surprising that the most advanced economies with the highest levels of per capita income and the most sophisticated industrial and information technology should also be prominent in services trade. Innovations in services have played a fundamental part in assisting the advancement of these economies.

More than 30 per cent ($114.5b) of world exports of services are included in the 'other services' group. While the United States heads the list for shipping, travel, passenger services and other transportation, it is only ranked fourth (6.7 per cent) for 'other services'. Top of the list is France (13.8 per cent of the world total), followed by the United Kingdom (10.2 per cent) and Germany (Federal Republic) (10.0 per cent). This suggests that there are significant variations in the kinds of services exported by different countries, especially for the 'other services' group. National balance of payments statistics (UNCTAD, 1989a, annex table 6) indeed show such variations. Thus in 1985 France generated large surpluses on trade in, for example, construction engineering ($1.3b), technical cooperation ($1.5b) and management services ($0.6b), whereas the

United Kingdom had surpluses in items such as commissions and brokerage ($2.0b), banking ($2.0b), insurance ($1.9b) and consulting and technical cooperation ($1.6b). Construction engineering ($0.8b) is a deficit item for the UK, and patents and royalties ($0.5b) for France. Similar examples, but involving surpluses and deficits on trade in a different mix of services, can be cited for many of the other countries shown in table 3.2. The common denominator for all the services involved is that most are producer services (including the shipping and other transportation trade mentioned earlier). These are the only services in which trade has shown a relative increase during the 1980s, accounting for some 5 per cent of world current transactions (UNCTAD, 1989a; GATT, 1989). Otherwise world trade in all services has remained relatively static even though the contribution of goods to world current accounts had declined from 73 per cent in 1969 to 68 per cent in 1986 (UNCTAD, 1989a). The balance has shifted largely as a result of dramatic increases in interest payments, also inflated by the expansion of financial intermediation in services trade during the 1980s. The slow growth of services trade also reflects the difficulty of trading services by comparison with supplying them by FDI. By the late 1980s the share of services in the world stock of FDI was close to 50 per cent and they accounted for 55–60 per cent of world flows (UNCTC, 1991).

Three groups of countries engage in international trade in services (UNCTAD, 1989a). The first group contains these countries with large deficits on trade in services; high interest payments on international loans to these countries are largely responsible for the deficit. Indonesia, the Ivory Coast, and several Latin American countries including Brazil are in this group. A number are major exporters of commodities and have a strong relative position in goods which they need to protect from imports but want to promote for export in order to attract the foreign exchange to service their large debts. A second group of less developed countries generates large revenues from supplying labour to other countries (wages being returned to dependants in the home country) and from travel, but are relatively weak in relation to trade in other services and in goods. Many of the non-petroleum-exporting countries in North Africa, the Caribbean and West Asia are included in this group. Tourism is an important source of revenue, as is the income from nationals employed in some of the adjacent oil-producing countries. The third and smallest group has strengths in services trade that have compensated for long-term deficits in goods trade. The Republic of Korea and Singapore are the best examples: both have also been strengthening their manufacturing base in response to demand from developed countries for competitively priced goods, while at the same time recognizing the scope for supplying some of the more specialized services to other less developed countries.

Services trade by developed countries involves a much wider range of activities, with particular strengths in knowledge-intensive producer services. Breadth is absent in the profiles for most of the less developed countries, which are largely engaged in trade in the form of tourism or the provision of services abroad by nationals. Both of these involve the movement of persons across national frontiers. Embodied services and factor services trade is concentrated in developed economies; trade in all services usually makes a much more positive contribution to the balance of payments. For most developing countries, trade in services is likely to be a deficit item. This is not to suggest that services as a proportion of trade in a number of developing countries have not grown dramatically (GATT, 1989). In India's case the share has risen from 13 to 23 per cent and for Egypt the figure has risen to 53 per cent as a result of its Suez Canal earnings.

The Role of Comparative Advantage

The laws of national comparative advantage (embodied in the Hecksher-Ohlin-Samuelson (HOS) model) have long been used to account for the patterns of trade in tangible goods. Endowments such as strategic location or raw materials are considered important, and comparative cost advantage is explained in terms of the relative abundance of factor inputs (Nusbaumer, 1987; Richardson, 1987; Hindley and Smith, 1984). But finance capital, political and cultural factors, and the characteristics of human capital have always been very low in the list of factors considered to be influential. These are more dynamic, less fixed national endowments and, although continuing to be evaluated relative to trade in goods, they are actually much more relevant for explaining trade in services.

Comparative advantage in services is derived from factors such as the pattern (and the level) of existing development in a country, including the degree of investment in education, research and technological development, and the extent of government regulation of services. These are, however, more difficult to define and to measure as inputs to be included in the HOS model. Consequently, some observers argue that the model has limited value for explaining the structure and dynamics of international trade in services (Petit, 1987).

Riddle (1986) distinguishes between cultural and economic comparative advantage. The latter is the more commonly accepted form of comparative advantage, but a country may also be able to utilize a particular cultural trait at the expense of competitors. An example is the adoption by the world's major airlines of the 'etiquette schools' used by Japan Airlines to train its customer service personnel (Riddle, 1986). Conversely, a country may be able to derive cultural comparative advantage because the service provided by some other country with a different culture is not compatible. Services sourced from countries with an extensive colonial history, such as France, Britain or the Netherlands, may also possess a form of cultural comparative advantage over countries that have not had administrative or trading ties with host countries before.

Economic comparative advantage is derived from physical capital (both environmental and man-made) and from human capital. The relative importance of these at any one time varies according to the type of service industry and the level of economic development. Human capital is a major source of comparative advantage for producer services and currently tends to favour developed market countries, while physical capital is a more significant determinant of the international trade potential of the developing countries. The importance of tourism in the service trade profile of many of the smaller developing countries is heavily dependent on the quality of their physical resource endowments: they may be unique, unpolluted, uncongested and exclusive because of the costs of getting to them. This is one way in which location can contribute to comparative advantage; it is also significant in instances where human capital is the source of comparative advantage. Time zone location – and therefore the ability to maximize trading time for international capital, financial and currency transactions – has been very important for the relative growth of global financial centres, notably New York, London and Tokyo. Nusbaumer (1987) urges caution, however, about over-enthusiastic application of the principles of comparative advantage to trade in services: since some services require very specialized knowledge, they are exchanged on the basis of absolute rather than comparative advantage (see also Richardson, 1988). Other services can, however, be readily standardized to meet the needs of final consumers or intermediate users, and trade will be determined by comparative advantage in the primary factor endowments rather than by specialized knowledge.

The principles of comparative advantage also apply to service MNEs. Gibbs and Hayashi (1989) list several major factors that can contribute to the ability of service suppliers to become more competitive. These include financial capacity; the ability to make effective use of telecommunications and information technologies; accumulated knowledge, skills and reputation; networking ability; established relationships between suppliers and buyers; presence in major markets; ability to offer a package of services; domestic market size; and appropriate government incentives. Two of the factors in this list are particularly important for the development of services trade by MNEs and by developing countries: telecommunications and human resources (knowledge, skills and reputation). Investment in telecommunications infrastructure is a strategic decision for developing countries since it meets public and private sector requirements, determines access to information networks and the ability to provide services, and is also a component of national security (Pipe, 1989; Bressand, 1989). Existing telecommunications services in most developing countries are highly regulated, nationalized facilities, but more open systems achieved by a combination of deregulation and investment in infrastructure will be necessary if information and communication-intensive services such as transport, insurance, banking and business services are to achieve any competitive advantage outside the domestic economy. But while making national telecommunications services more open and improving the scope of domestic suppliers to capitalize on any comparative advantage, a developing country also becomes more accessible to foreign MNEs and therefore more vulnerable to global economic trends and business cycles. National sovereignty and territoriality are also compromised and some developing countries remain uncertain about the costs and benefits of seeking greater participation in the international market for services (see for example UNCTAD, 1989b). It is certainly the case, however, that service MNEs cannot seriously consider establishing (either to import or to export) in developing countries that have inadequate telecommunications infrastructure; they must either be prepared to invest in their own intracorporate networks (which is unlikely to be acceptable to the host country) or be persuaded by the host country to invest in its telecommunications infrastructure as a condition of receiving permission or a licence to establish a local presence.

The relationship between human resource development and comparative advantage has become crucial in recent years. It has been explored by Bertrand and Noyelle (1986; 1988; see also Noyelle and Stanback, 1988) in a study of changing technology, skills and skill formation in financial services firms in five countries. Financial services firms, for example, are relying less on internal, vertical labour markets (the traditional structure) and more on external, horizontal labour markets involving greater interfirm labour mobility, more specialized skills and a more frequent need for upskilling. Because clients are becoming more demanding, financial services workers need better interactional skills (including languages) and the distribution and marketing of services are more important than mere production. Individual workers must be not only flexible but also able to identify and solve problems at whatever organizational level they operate. Some of these skills can only be acquired by externalizing their production outside financial services firms; others are provided by vastly increased investment in internal, firm-based training. All these factors, and others not enumerated here, point to the need for countries (developed and less developed) to raise the overall level of educational attainment of their labour forces. Bertrand and Noyelle (1986) suggest that footloose service MNEs will increasingly include assessments of the variations in human resource qualities between countries in their strategic development plans; it is therefore vital that countries with existing comparative advantages in this area for selected services do not become complacent. The task for other countries is to try to narrow the gap by adopting positive policies for upgrading human resource skills.

Table 3.3   Estimated service sales for US industries in overseas markets and non-US-based industries in the US market, 1974

US industries From the US	$b	Non-US industries To the US	$b
Total foreign sales *of which:*	50.0	Total sales to US *of which:*	28.8
Exports	7.0	US imports	7.9

Overseas affiliate sales
  in:
	43.0	Affiliate sales in US
		  in:
			20.9

Banking	12.0	Banking	4.4
Insurance/other finance	2.1	Insurance/other finance	6.9
Wholesale/retail trade	6.4	Wholesale/retail trade	6.6
Advertising	3.4	Advertising	0.1
Franchising	1.5	Franchising	–

Transport,
    communication and
    utilities
	2.7	Transport
		    communication and
		    utilities
			1.8

Other	14.9	Other	1.1

Service sales as % of
  goods sales
	12.1	Service sales as % of
		  goods sales
			24.8

Source: US Department of Commerce, 1976

Foreign Direct Investment in Services

Firm-specific or competitive advantage may be measured in a number of ways, but firms must still identify the best way to actually service overseas markets. The choices they make will be partially governed by the regulatory environments that control trade in specified services or barriers that prevent entry to national markets. For many services the effect is to discourage exports but to encourage the supply of a service through direct overseas production. Indeed, for many services that embody knowledge or expertise in people there is no choice but to be directly represented overseas (unless it is judged acceptable to grant a licence to a third party to provide the service, but this may cause loss of control over the knowledge contained in a product, the quality, the price or all these things). It certainly appears from US evidence (table 3.3) that service industries prefer overseas production to exporting from their home base. Over 80 per cent of total foreign sales by US service industries were derived from overseas affiliates; a similar feature is apparent for sales of services by foreign service firms in the US. An important reason for direct production rather than export is that information-intensive services, such as banking businesses or management consulting services, can be more profitably distributed from within the same organization. This also reduces the risk of expropriation of specialist knowledge or skills.

Foreign direct investment is therefore the best way for firms to maintain their competitive advantages across national boundaries. A growing number of companies have expanded across borders by making foreign acquisitions, merging with competitors and investing in greenfield sites. FDI has now reached a level where it is contributing to fundamental changes in the structure of the world economy and creating a qualitatively different set of linkages between advanced economies (DeAnne, 1990). It increased at more than 20 per cent per annum between 1983 and 1988, and the worldwide stock of direct investments by the G5 countries (US, UK, Japan, West Germany, France) totalled $757b. The growth in FDI is increasingly concentrated in developed countries; flows to developing countries fell in real terms during the 1980s. Furthermore, 75 per cent of the world's FDI stock is held by the G5 countries, but they account for only 42 per cent of world trade (see also UNCTC, 1991). This differential has been occurring at a time when the most important factor contributing to FDI growth has been international liberalization of services, especially during the second half of the 1980s (table 3.4). The regulatory structures that have protected many services behind national boundaries are beginning to break up, and this process will escalate as countries try to protect their competitive advantage, especially for international service activities. Service multinationals are the key actors and beneficiaries. Telecommunications operators such as AT&T and British Telecom have now started to engage in major cross-border acquisitions or in joint ventures (table 3.5). Some 25 cross-border acquisitions were made in 1988 (value $2.1b) compared with only six in the previous year (value $120m) (Booz-Allen and Hamilton, 1990). This is a precursor to accelerated expansion as liberalization of national telecommunications services continues.

It is difficult to disaggregate the composition of services FDI, but where it is possible the 'similarities are nevertheless more striking than the differences' (Sauvant and Zimney, 1989, 80). The growth

Table 3.4   Stock of FDI investment in services, selected host developed and developing countries, various years

Country and currency	Year	Value Total FDI	FDI in services	Services as % of total FDI
Developed countries				
Japan ($b)	1975	1.5	0.3	18
	1983	4.9	1.2	25
United States ($b)	1974	26.5	11.5	43
	1985	182.9	92.2	50
United Kingdom (£b)	1971	5.6	0.6	11
	1981	30.0	7.3	24
France (francs b)	1980	89.7	33.1	62
	1985	129.0	81.7	63
West Germany (DM b)	1976	78.9	25.9	33
	1985	119.2	57.1	48
Developing countries				
Mexico ($b)	1971	3.0	0.6	19
	1981	13.5	3.2	23
Brazil ($b)	1971	2.9	0.5	16
	1985	25.7	5.6	22
R. of Korea ($b)	1980	1.1	0.26	23
	1986	2.1	0.56	27
Thailand (bhat b)	1975	10.3	5.7	55
	1984	48.4	23.0	48
Morocco (DH b)	1975	0.8	0.4	48
	1982	4.5	2.4	54

Source: extracted from Sauvant and Zimney, 1989, 107–110, table 2 and table 3

has been taking place mainly in financial services and trade together with the proliferation of finance affiliates established by trading, insurance, manufacturing and oil company TNCs from the developed countries. Funding has traditionally been dominated by Western Europe and US retailers, commodity traders and Japanese *sogo shosha* (Sauvant and Zimney, 1989; Dicken, 1992). The last are massive conglomerates spanning commercial, industrial and financial interests whose main activity is the organization of trade: they operate through a complex web of subsidiaries scattered around the world. *Sogo shosha* are major players in the global economy and in the service economy in particular: in the second half of the 1980s they were involved in 8 per cent of world trade and as much as 17 per cent of the trade of Asian and Pacific countries (UNCTC, 1988, cited

Table 3.5   Major cross-border acquisitions, investments and joint ventures, telecommunications operators, 1988–1989

AT&T (US)	Istel (UK)
	Italtel (Italy)
	Network Systems Int. (Netherlands)
	Olivetti (Italy)
Bell (Canada)	Cable TV (UK)
Bell South (US)	Air Call communications (UK)
British Telecom (UK)	McCaw Cellular (US)
	Mitel (Canada)
	Tymnet (US)
Infonet (US)	Belgium PTT
	Deutsche Bundespost
	France Telecom
	Netherlands PTT
	Singapore PTT
	Telecom Australia
	Telefoncia (Spain)
	Teleinves (Sweden)
Telefoncia (Spain)	Entel (Chile)

Source: Booze-Allen and Hamilton, 1990, after Dixon, 1990

in Dicken, 1992). Mitsui and Mitsubishi, for example, were amongst the top ten corporations in the world in 1985 with combined sales of almost £100b. But the share of services FDI has been boosted by the expansion of marketing and sales affiliate networks by developed country manufacturing corporations. A certain amount of these services cannot be supplied outside the home country without a large investment in fixed assets; hence the average FDI per US affiliate abroad in banking, insurance and retailing in 1982 was $12m, whereas in other services such as advertising or accounting it was only $3m (US Department of Commerce, 1985).

The degree of transnationalization through FDI varies across the service industries. On the basis of evidence from US service transnationals Sauvant and Zimney (1989) identify three groups. Firstly, there are services where FDI is the exclusive or major way of delivering to foreign markets, e.g. advertising, leasing, investment banking, securities broking, accounting, insurance, engineering, data processing and retail trading. Secondly, there are services where exports are the main method of delivery, e.g. franchising and licensing, education and legal services, and travel. Thirdly, for some services both FDI and exports are important, e.g. construction, film rentals, health, information, consulting, software, transportation and communications (see Vandermerwe and Chadwick, 1989).

Flows of FDI have increased dramatically during the 1980s, with a distinct shift in sectoral composition towards services and high-technology manufacturing (UNCTC, 1989; de Smidt, 1992). This sectoral shift to some extent reflects the changing composition of national employment or GDP in many countries (see chapter 1). It also mirrors the decline of FDI in many resource-based activities such as petroleum as industrial countries have nationalized ownership of production that was formerly undertaken by trans-national corporations (UNCTC, 1991). The geographical patterns of FDI in services are largely determined by the same group of countries that are prominent in international trade (see table 3.2). The UNCTC has produced estimates for the composition of the stock of outward FDI by six of the leading countries in 1975 and 1987 together with the compound annual growth rate for each sector (table 3.6). FDI has been growing in all three sectors, but in all six countries the highest rate of growth has been in services FDI and always at a rate above the average for total investment. The European Statistical Office (Eurostat) has shown how the creation of the Single European Market has stimulated investment by trans-national service corporations; the share of services in total investment in the EC by Community transnational corporations as well as by countries outside the EC rarely fell below 60 per cent between 1984 and 1988 (UNCTC, 1991).

Further sectoral shifts towards services FDI and continuing growth in the value of investments can be anticipated during the 1990s. An important factor will be the demand for telecommunications, financial accounting, legal and banking services generated by the transition to a market economy in the countries of Eastern and Central Europe (see below). The demand for advanced producer services will continue to expand in developed and developing economies as modernization and restructuring of the production process continue. Removal of the regulations governing flows of FDI is only just beginning but is spreading to more countries and includes a wider range of services such as public utilities,

Table 3.6 Sectoral shares and growth rates for outward FDI of major home countries, 1975–1989

Country	Year	Sector Primary	Sector Secondary	Sector Tertiary	Total
Canada	1975[a]	21.1	50.5	28.4	100
	1987[a]	13.1	43.3	43.4	100
	1975–87[b]	12.1	15.1	16.6	16.4
France[c]	1975	22.1	38.2	39.7	100
	1988	15.0	36.6	48.3	100
	1975–88	22.6	26.0	28.3	26.3
Germany	1976	4.5	48.3	47.2	100
	1988	2.8	43.4	53.7	100
	1976–88	7.6	10.7	12.9	11.5
Japan	1975	28.1	32.4	39.5	100
	1989	6.7	26.0	67.0	100
	1975–89	10.0	19.9	26.5	21.8
Netherlands	1975	46.8	38.6	14.6	100
	1988	36.4	24.7	38.8	100
	1975–88	5.9	4.2	16.5	7.9
United States	1975	26.4	45.0	28.6	100
	1989	16.7	40.9	42.3	100
	1975–89	4.7	7.4	11.4	8.2

[a] Percentage shares.
[b] Compound annual growth rates.
[c] Based on cumulative flows of direct investment from 1972.
Source: UNCTC, 1991, 17, table 7

telecommunications and transport which have traditionally been protected from foreign ownership. The increase in transborder data flows stimulated by technological change (chapter 2) may also stimulate FDI as service TNCs (following the behaviour of manufacturing TNCs) disaggregate their operations in order for example to take advantage of lower usage costs or differentials in accommodation costs in different global locations.

International Trade in Services and the Developing Countries

Closer examination of the IMF balance of payments statistics shows that exports of all services from the top twenty developing countries

Table 3.7   Major exporters and importers of all services, developing countries, 1984

Country	All services, exports ($b)	(% other services)	Rank	Country	All services, imports ($b)	(% other services)
Singapore	9.7	28	1	Saudi Arabia	14.2	64
R. of Korea	6.4	44	2	Mexico	5.4	22
Mexico	5.8	35	3	R. of Korea	5.3	32
Saudi Arabia	4.4	93	4	Brazil	4.9	20
Egypt	4.3	16	5	Malaysia	4.6	30
Yugoslavia	3.7	22	6	Indonesia	4.5	36
India	3.2	63	7	Yugoslavia	4.2	48
Brazil	2.3	22	8	India	4.1	22
Turkey	2.2	41	9	Singapore	4.0	48
Malaysia	2.2	23	10	Kuwait	3.2	–
Thailand	2.0	10	11	Venezuela	3.2	13
Argentina	1.8	–	12	Egypt	3.0	43
Panama	1.5	–	13	Iran	2.8	–
Philippines	1.4	64	14	Algeria	2.4	38
Jordan	1.2	17	15	Argentina	2.1	–
Bahamas	1.1	–	16	Libya	1.9	–
Neth. Antilles	1.1	–	17	Thailand	1.8	–
Columbia	1.0	20	18	Nigeria	1.5	33
Venezuela	1.0	–	19	Columbia	1.5	20
Kuwait	1.0	–	20	Chile	1.4	–
Top 20 countries	57.3	35			76.0	35
Total developing countries	75.6	31			101.6	32

Source: extracted from UNCTAD, 1989a, 291, annex table 13, and 292, annex table 14

amount to less than 20 per cent of the total for the top twenty developed market economies (table 3.7; see also table 3.2). The figure for the latter does, however, include Singapore, Republic of Korea, Mexico and Saudi Arabia, and together these account for over 45 per cent of total service exports by the top twenty developing countries. There are slight variations in the rank orders for exports in shipping, travel, passenger services and other transportation. But the 'other services' exports (35 per cent by value of all service exports by the top twenty countries) are firmly dominated by these four countries (57 per cent). As might be expected, developing country imports of services exceed exports by a margin of some 34 per cent (compared with 6 per cent for developed countries), although the total value of imports is only 25 per cent of the figure for developed countries. Imports of 'other services' account for some 34 per cent of the total and are purchased primarily by Saudi Arabia, Yugoslavia, Singapore and the Republic of Korea.

International trade in services generally presents developing countries with many more problems than the developed countries. For the latter the symbiosis between manufacturing growth and services growth and, latterly, the emergence of producer services have been capitalized upon by internationalization, the trend by many corporate organizations towards externalization of service production, and their ability to use the latest technology to develop information-related services. The resulting comparative and competitive advantage is considerable and they not only export large quantities of services but view them as making an important and increasing contribution to national current accounts. Trade surpluses in services for countries such as France and the UK help to compensate for deficits on trade in goods. Deficits on services trade are more normal for developing countries, with tourism being the main exception. Mexico provides an example of the problems facing many developing countries, although it is important to stress that it is difficult to generalize across the experience of all the developing countries (UNCTAD, 1989a).

Mexico

Mexico's key problem is meeting its debts: net outflows due to interest payments, transportation and property income were larger in the early 1980s, for example, than the surpluses on merchandise trade. Service exports can help to rectify the trade balance: tourism is a major export earner but its contribution has tended to be irregular owing to fluctuating environmental and political circumstances. Processing activities (*maquiladoras*), whereby international subcontracting work is undertaken in Mexico, is a second major source of service revenues. The third source of services revenue is 'other services', especially reinsurance and recent improvements in the competitive position of Mexico in architecture, construction engineering and computer services. Much of this activity is directed at Mexico's trade links with other Latin American countries, although a large proportion of its tourism revenues comes from US visitors. But while it is trying to diversify its service exports, Mexico must import services to support the process of industrial restructuring that will make its manufactured goods more competitive. Other services, especially producer services, therefore comprise a large part of Mexico's service imports, which amount to 30–40 per cent of the total.

The ability of Mexico to increase and diversify its trade in services also depends on efforts to improve the role of services in its domestic economy. With services accounting for some 60 per cent of GDP in 1984 (UNCTAD, 1989a) Mexico does not, on the surface, seem to be in a position very different from that of many developed market economies. However, closer examination of the structure of the Mexican service sector shows that: financial and business services make a very small contribution to GDP (less than 4 per cent); there is a high concentration of relatively low value-added commercial service activities; and a high share of revenues comes from real estate. A Mexican government study (de Mateo, 1988) set up to explore the relationship between services and other parts of the national economy and ways of stimulating the growth of key services notes: a decrease in service sector productivity (1970–84); the constant share of services in GDP during the early 1980s; a tendency towards internationalization of service production within manufacturing (contrary to the externalization trend by manufacturing in developed countries); the inadequacies of the infrastructure (telecommunications in particular) required by information services; and the importance of fostering a closer relationship between industrial restructuring and producer services (as well as services in general) if the Mexican economy is to grow on the basis of external trade in merchandise and in services.

India

The services share of GDP in India and its contribution to credits in the balance of payments (1980–85) has been increasing (see for example Riddle, 1989). The balance is most favourable for exports in the 'other services' category which, as in other developing countries, is the most dynamic component of services trade. But this is precisely that part of services trade that is most difficult to enumerate. It is therefore necessary to identify specific services in which a country like India has (or may have) comparative advantage and to conduct case studies of their behaviour and growth (UNCTAD, 1989b). Computer services (world market worth $55b in 1985, estimated $341b in 1996) is one activity in which India may be able to rapidly develop its exports (Srivastava, 1989). Exports of computer software from India were valued at Rs30m in 1980 but are anticipated to reach more than Rs3b in 1991. The information processing industry is growing at 40 per cent per annum with software comprising 40–45 per cent of the market. In recent years the expansion of information processing industries has depended much more on advances in software than in hardware. Indeed, hardware costs have been falling dramatically while software costs have continued to rise. The reason for the differential is the labour-intensive character of software development combined with an apparently insatiable demand for new, more powerful and faster programs for an ever widening range of applications. As a consequence a gap has opened between the demand and supply of labour skilled in software development, further escalating its cost relative to hardware in information processing. India has the third largest pool of scientific and technological manpower in the world (Srivastava, 1989), which is also English speaking and is therefore uniquely positioned to become a major international software supplier. Packaged software (50 per cent of the world market in 1985) has the best potential for growth but is the most risky and involves the highest initial investment. This leaves professional services as the area in which India might target its software production during the 1990s. Most of the software from India comes under this heading with 80 per cent provided on site to overseas clients. This is an expensive and potentially limiting way of achieving growth in this market because host countries either have existing barriers to entry of overseas professionals or will introduce them if India is seen as a threat to domestic software producers. Thus, India must seek to increase the proportion of overseas work conducted in India; this will require improved telecommunications and better marketing.

Developing Country Service MNEs

Given that the distinction between exports of services and the activities of multinational service firms is, for much of the time, difficult to make, the role of developing country service MNEs in international trade should not be overlooked. But, as Lecraw (1989) suggests, most developing country service MNEs invest in relatively inexpensive human capital for their overseas operations rather than in capital, plant or equipment because foreign exchange at home is usually in short supply (and there are often controls on the export of finance capital) while labour is abundant. Their capital needs abroad are obtained through ventures or in host country capital markets. Lecraw (1989, 204) defines developing country service MNEs as 'firms that transfer capital (human, money and physical) *within the boundaries of the firm*. The important characteristic is that resource transfers are made *within* the firm, not via the market'. A firm from a developing country that used its own workers to undertake construction projects abroad would be a service MNE, while a developing country firm that subcontracted its workers to and for another construction firm in the host country would be engaged in exporting labour services. In shipping, banking, hotels and construction, developing country multinationals are now significant: 30 per cent of all employees in foreign banks in New York are working for banks with home offices in low- and middle-income countries (Lecraw, 1989). Large hotel groups operate throughout Asia (e.g. the Regent Group) and Korean construction firms employed 60,000 Koreans in the Middle East during the late 1970s. Significant comparative advantage in services of this kind (often including leadership in service technology) contrasts with the almost total absence of developing country service multinationals operating in media services, financial services such as futures, brokerage and currency trading, health care, telecommunications and retail distribution. Developed country service MNEs have the competitive advantage in these services.

Eastern Europe, the Commonwealth of Independent States and International Trade in Services

There can be little doubt that as the accelerating process of reform (*perestroika*) continues in Eastern Europe and the former USSR (the new Commonwealth of Independent States) the service industries will come under increasing scrutiny as a result of irresistible demand for internal economic restructuring and external competition. As countries such as Hungary, Russia, the Ukraine, Poland and Czechoslovakia endeavour to initiate market-oriented economic systems (as distinct from their centrally planned, highly bureaucratic predecessors) they will need a new approach to the supply of consumer services and, most important, to the producer service industries that support production (Maciejewicz and Monkiewicz, 1989; Bouska and Cerny, 1991). The development of international trade in services will be central to the process of adjustment; but in Czechoslovakia, for example, the principal service export is transport (railways and pipelines) rather than business services. Even if the institutional obstacles to trade in these services are removed it will still be necessary, at least in the short and medium term, to import these services while developing a domestic base which will involve the direct participation of foreign capital. According to Bouska and Cerny (1991) the prospects are encouraging: one-third of foreign direct investment into the Czechoslovak economy during the first nine months of 1990 went into services. This flow will increase as licensing procedures or problems of establishing property rights are simplified. The technology gap that already exists relative to developed and some developing countries will increase even further if certain types of services, notably telecommunications-related activities, are not developed as a matter of priority (this includes extensive and intensive modernization of existing facilities and investment in new infrastructure). Financial services were never well developed in most socialist economies so that financial instruments and loan facilities for venture capital, for example, are very limited. Typically, the former socialist countries have a national state bank, an investment bank and a specially created foreign trade bank to provide finance for international transactions (Hill, 1989). Some countries, such as Hungary, have responded by introducing new banks in direct competition with their national (state) bank and allowing co-ownership of their banks by foreign banks. The latter will encourage exchange of badly needed financial expertise, especially in the operation of international markets. Hungary opened Eastern Europe's first stock exchange in mid 1990. With a daily trading average of £200,000 the exchange, which is open for one hour each day, is extremely small by world standards. Warsaw's stock exchange was also reopened in April 1991 after being closed for 50 years.

Recent levels of trade in services by the countries of Eastern Europe have therefore been low – equivalent to some 5 per cent of world trade in services in 1984 (UNCTAD, 1989a). Estimates (apart from the IMF balance of payments data for Romania, Hungary and Poland) show that the former socialist countries as a whole, in contrast to the situation for the developing countries, register a surplus in their net service trade with market economies. This is an important factor in financing the trade deficits of the countries that comprised the former USSR, East Germany and Poland. Because many of the developed market economies have, until very recently, placed tight embargoes on trade in many goods (especially those embodying high technology) with the former socialist countries, they tend to be more important trade partners in services. On the other hand, trade in services between the former socialist states has been insignificant. Their services exports are largely confined to other services, shipping and travel; shipping also dominates imports along with other services such as assembly engineering and construction. More advanced services such as advertising, management consulting, insurance, and licences and patents have largely been absent from services trade by the former socialist countries. This is a pattern similar to that for developing countries, and will be the segment of international trade where they have the most 'catching up' to do if they are to successfully adopt a more open approach to the organization of their economic systems during the 1990s. They can no longer afford to subordinate the development of the service sector to the production of goods.

Transport and communication and other services contributed to some 90 per cent of Poland's export earnings in 1986; they were also responsible for 85 per cent of the country's import expenditures on services (UNCTAD, 1989a). Professional and technical services represented one-third of the positive balance, with much of the trade taking place with developed market economies, followed by other socialist countries. Trade in services with developing countries was of minor significance, but the balance has been shifting in recent years towards greater interaction with countries such as Czechoslovakia and Hungary and a diminishing level of trade with both developing and developed countries. Maritime transport, which has been allowed to function in a more liberal and competitive way than many other Polish service activities, is the principal service category in international trade (for both imports and exports). Poland exports more professional services such as specialist consulting and construction engineering services than it imports, but structural difficulties in the domestic economy and barriers to trade in international markets mean that this sector's contribution to Poland's balance of payments is under-represented. There are clearly parallels with the circumstances confronting most of the developing countries.

It is likely that service multinationals controlled from Poland and elsewhere are making a modest contribution to international trade that is not adequately recorded in balance of payments data. Some evidence based on the activities of British and Swedish subsidiaries of Soviet and East European service multinationals has been collected by Hill (1989) (see also Knirsch, 1983; McMillan, 1979). This shows that most of the subsidiaries are providing marketing support for the sale of goods manufactured in the home country. They also act as agents for home country establishments wishing to purchase components etc. required for domestic production. Some of the subsidiaries are also engaged in transport, insurance and banking services, but the scope for expanding these activities is restricted by the scarcity of suitable expertise and legal awareness in the home country and the scarcity of foreign exchange (arising from international trade in general) to support further foreign direct investment. Thus, with some exceptions such as the Moscow-Norodny Bank, the scale of operation of new Commonwealth and East European subsidiaries has been very small by comparison with market economy banks and other MNEs. This situation will now change quite quickly as barriers to trade between East and West are removed as a result of the political and economic changes that have been taking place since late 1989, culminating at the time of writing with the dissolution of the USSR and the creation of the new Commonwealth.

Liberalizing International Trade in Services

It has long been recognized that one of the motivations for international trade in goods is that a number of benefits accrue to the countries and corporate organizations that are involved. Advantages arise from the exploitation of economies of scale, from the incentives to become more competitive in order to retain market share, from the need to improve the design and quality of products for much larger and more diverse markets, from allowing specialization in the goods and products in which countries have expertise, and from the downward pressure on prices from intermediate and final consumers. There is no reason to believe that such benefits should not arise from international trade in services. Services trade in a more open system will, for example, stimulate competition that would diversify and improve the quality of services available in different markets. It will encourage domestic service firms to become more efficient in the face of competition from outside firms, and service production will be reallocated in a more efficient way between countries and markets. As a general rule, industrialized countries have a comparative advantage in technology- and capital-intensive services; developing countries have a comparative advantage in labour-intensive services.

The challenge for developing countries is how, within a multinational framework, to provide for the export of these services to industrialized countries. But, by comparison with merchandise trade, efforts to stimulate services trade through bilateral or multilateral negotiations involving individual countries or international organizations have advanced at a snail's pace. This is true even though, as has already been seen, trade in goods is a stimulus to trade in services as well as an activity that can increasingly take place independent of trade in goods. Yet most countries are not convinced that services trade is important to the advancement of their economic development. Some of the reasons will be apparent from the earlier discussion of the extensive measurement and identification problems: negotiators must work without objective evidence about the existing attributes of the system that they are endeavouring to stimulate.

Moves to introduce freer trade in services were started by the US during the 1970s but met with little success (Feketekuty, 1988; Cowell, 1983). Along with other industrialized (G7) countries such as the UK and France, the US stood to benefit from free trade; the less developed countries were more sanguine, however, and have continued to be suspicious about trade which, it is anticipated, will shift comparative advantage even further towards the developed countries. The US takes the view that there is an ever expanding world market for goods and services; foreign competition will stimulate domestic markets and assist the balance of trade in services. The early moves to promote liberalization were also taking place when it was unclear which international organization would provide the best 'umbrella' for multinational negotiations or whether liberalization should be directed at services trade in general or through a sector by sector approach.

Clear answers to these questions remain elusive even though the Uruguay Round of negotiations, under the auspices of the General Agreement on Tariffs and Trade (GATT), include a commitment to devising a multilateral framework for trade in services. It is intended that this should lead to an agreement by all countries to the progressive liberalization of trade in services in a way that satisfies those countries already in a position to benefit immediately and which is also consistent with the growth strategies of developing countries and the expansion of their service exports. The basic aim is to remove those aspects of national regulations which discriminate against foreign suppliers. This is not the same as deregulation, which determines the way in which individual service markets function within countries; in some cases liberalization will make it necessary to make national regulations, in financial services for example, more rather than less rigorous.

Riddle (1986) suggests a framework that might be used to approach negotiations in service trade (table 3.8). Rather than focusing on sectors, it is suggested that more attention is given to the different ways in which services are delivered with respect to the factors of production and to consumers. There are many more alternatives than those available for merchandise trade, where it is assumed that factors of production are fixed at one or more locations

Table 3.8   Model of alternative services trade strategies

	Factors of production:	
	Don't move	Move
Consumers:		
Don't move	Cross-border trade	Foreign earnings trade
Move	Domestic establishment trade	Third-country trade

Source: Riddle, 1986, 196, table 9.1

and the merchandise is moved to the consumer. Depending on whether consumers and the factors of production do or do not move, there are four strategies for services trade; only one of these is similar to merchandise trade. Cross-border trade, such as that engaged in by communications, professional services or transportation, does not require the movement of factors of production or consumers. But this does not make measurement of this trade any easier because the telephone and postal systems are extensively used by professional services. Foreign earnings trade requires the factors of production to move but not the consumers. Therefore countries that export their labour services and whose earnings are repatriated would be an example of cross-border trade. Individuals (customers) may also receive income from investment in another country without actually owning any enterprise, e.g. returns from unit or investment trusts that incorporate in total all or in part the shares of overseas companies. The repatriated earnings of foreign affiliates are also included under foreign earnings trade.

The other two trade strategy alternatives are domestic establishment and third-country trade (table 3.8). The former involves movement of the consumer but not the factors of production; examples are specialized health care, higher education and tourism. This is an extremely important form of services trade for many countries, especially in the developing world. Such countries are also interested in, and concerned about, third-country trade whereby tourists from another country may patronize services such as hotels, fast-food services or entertainment that are provided by multinational enterprises, the revenues being repatriated to the home country of the multinational. More widespread is the purchase of business services by a foreign bank branch in any of the world's major cities which are provided by service multinationals in management consulting or advertising, for example. Although the classifications suggested in table 3.8 may help individual countries to target the most appropriate trade strategy, it may well be the case that more than one strategy is important and they will not have the data available to assist identification of priorities. Since such priorities will inevitably differ between countries, negotiations at the multinational level will have to fall back on a process of liberalization aimed at reducing the worst effects of laws, regulations and administrative practices on services trade.

We have seen that regulation of services through the application of non-tariff barriers is pervasive (chapter 2). But the situation is complicated by the fact that multilateral agreements are already in place for some services but not at all for others (Gibbs and Hayashi, 1989). Thus, the International Air Transport Agreement (IATA) and the International Civil Aviation Organization (ICAO) incorporate almost all of the world's commercial airlines (some of which are state owned) and have multilateral regulations that are regularly monitored and updated when required. A similar situation applies to telecommunications in the form of the International Telecommunications Union (Pipe, 1989) and to shipping which operates under the multilateral regulatory framework provided by the United Nations Code of Conduct for Liner Conferences (Faust, 1989). The latter is a mechanism for equitable distribution of cargoes between groups of liners, ensuring a certain minimum share to the national shipping lines of developing countries. Because many of these agreements are industry specific and often also incorporate some bilateral arrangements (such as those between airlines on the North Atlantic routes between the UK and the US) there is scope for restricting competition to certain areas or facilitating international competition by ensuring lucrative domestic markets. This is clearly not in the spirit of liberalizing service trade. Much the same can be said of the frameworks for the liberalization of trade in services that already exist for the countries that are members of the Organization for Economic Co-operation and Development (OECD) (Everard, 1987). But the OECD comprises a limited number of member states and, although most accept the validity of instruments such as the Code on the Liberalization of Current Invisible Operations, many still maintain a variety of national measures that contradict the principles and objectives of these codes. By applying peer pressure, there is a gradual liberalization of these 'reservations' by individual countries; once removed they cannot be reinstated.

The most comprehensive prospects for liberalizing trade in services via multilateral negotiation and agreement rest with the GATT. Attempts to devise a multilateral framework have been prominent during the Uruguay Round of negotiations (these include trade in goods) following pressure from the US government. This was strongly resisted by the developing countries, notably India and Brazil, who consider this would give *carte blanche* to predatory multinational service companies, but a compromise was reached at Punta del Este (Uruguay) whereby trade in goods and trade in services are treated as separate agenda items so that concessions in goods trade cannot be traded for concessions on trade in services. A proposal to establish a special GATT for services alone is still being studied. The negotiations on services trade are complex and uncertainty about the outcome remains even though some progress has undoubtedly been made. It will be sufficient here to outline the key concepts, principles and rules that are being used to move towards a multilateral framework. The main considerations are: transparency, national treatment, most favoured nation status (non-discrimination), market access, increasing participation of developing countries, and progressive liberalization. Each of these is briefly summarized below.

Transparency refers to the availability of information about all laws, regulations, administrative guidelines and international 'agreements' relating to services trade to which the signatories are parties. The objective is to ensure that all participants are fully aware of their rights and obligations arising from trade-related rules; the problem is agreeing on whether individual rules are trade related or not. The developing countries, such as Mexico, take the view that if agreement on transparency cannot be reached (in relation to any particular regulation or rule) it should be excluded from the negotiations. National treatment is taken to mean that service exports and/or exporters of any category are accorded in the market of any signatory, in respect of all loans, regulations and administrative practices treatment no less favourable than that accorded to domestic services or service providers in the same market. This is one of the key agreements in the Canada–US Free Trade Agreement (see chapter 2). The importance of fair and equitable treatment is not contestable but the developing countries are anxious to ensure that it is applied only in the context of other provisions within the multilateral frameworks, thus avoiding the need for them to protect themselves or to regulate foreign direct investment because of the simultaneous application of the principle of market access (see below).

The concept of unconditional non-discrimination (or most favoured nation) has been important for the liberalization of trade in goods for more than 40 years. It means that the countries that are party to the GATT accept the broad multilateral concept of reciprocating, i.e. they agree to provide market access to trading partners in sectors where they do not possess an effective export capacity in return for compensating concessions with respect to sectors of export interest to them. In principle, there is no reason why non-discrimination should not be applied to trade in services; in practice, the developed and developing countries are some way apart, with the former wanting conditional most favoured nation treatment and the latter more interested in an unconditional arrangement. The anxiety is that equal application of non-discrimination cannot take place amongst unequal parties at different stages of development and whose service industries may be damaged rather than stimulated and strengthened by such liberalization. The principle of market access is also contentious but there is general agreement on two provisions: firstly, that it should be consistent with other provisions within any multilateral framework and, secondly, that the preferred mode of providing each service should be permitted. The main difficulty is how to distinguish between trade in services and investment, even though the latter may be necessary to a limited degree for the efficient delivery of, for example, insurance services in a host country. Developing countries are concerned that what the developed countries mean by market access is the 'right' of establishment that might ultimately allow them (or their MNEs) to become a monopoly supplier of specialist services.

The inclusion of services in the GATT has long been viewed by developing countries as likely to lead to a bias towards developed countries. At a meeting in Montreal in late 1988, trade members of the GATT member countries unequivocally stated that 'the framework should provide for the increasing participation of developing countries in world trade and for the expansion of their service exports, through the strengthening of their domestic service capacity, its efficiency and its competitiveness' (Modivel, 1989). One of the factors most likely to help achieve increased involvement of the developing countries will be the requirement, agreed in Montreal, for improving their access to distribution channels and information networks. It is believed that this will allow them to counter the success of the 'all-embracing' service multinationals, which is derived not only from their financial strength but also from their undoubted superiority with respect to delivering services using advanced information technology and information networks to which they often possess exclusive access (for a full discussion see Bressand and Nicolaidis, 1989; Gibbs and Hayashi, 1989). Finally, liberalization of services trade must take place progressively (OECD, 1990; Richardson, 1988). The development situation of participants in multilateral agreements varies, so flexibility of implementation should be allowed. National policy considerations relating to development objectives, especially in developing countries, would mean that the use of FDI, for example, as a method of providing services to these countries would not immediately qualify for any equal or liberalized status.

In a recent study by the OECD (1990) it is argued that developing countries stand to gain economically, in particular as a result of skill transfers, by agreeing to liberalize trade in services. On the basis of a sector-by-sector analysis the OECD suggests that most developing countries will have areas of export opportunity that would be better exploited or development opportunities that could be enhanced by improved access to imported services and the skills transfer with which they are frequently associated. The US Coalition of Service Industries (CSI) has conducted a survey of US service sector companies in developing countries to show how they have promoted development by creating additional employment opportunities, fostering technology transfer and boosting exports (CSI, 1989). Ten companies (in banking, telecommunications, construction, insurance and related services) had created 48,000 jobs in fifteen leading developing countries; 99 per cent were held by host country nationals and 83 per cent of management positions were held by host country nationals. The technology transfer involved in the expan-

sion of these companies led to higher exports. Examples given by the
CSI include advertising services developed in Brazil, Malaysia,
Hong Kong, Mexico, Singapore, Taiwan and Thailand for export to
surrounding countries; specialized computer software for small
bankers exported to the UK, the US, Malaysia, Philippines and
Indonesia; and Arabic computerized script developed in Kuwait for
export to other Arab-speaking countries. The result will be that even
the poorest nations may be able to develop a market niche at the
appropriate level of skill. Thus, the success of the Uruguay Round
in the area of labour mobility will be crucial; it will, for example,
allow developing countries to properly benefit from the opportunities for exporting professional services. Tunisia, for example, had
7000 engineers, doctors, teachers and other professionals working in
other areas of French-speaking Africa, replacing Europeans at one-
half to one-third of their cost.

The guiding principles for liberalizing trade in services will evolve
with the changing scale and strength of the world economy and the
extent to which the balance between the developed and less
developed countries changes. Meanwhile the actual process of
negotiation continues; this takes place in the context of the
provisions sought by participants relative to the objectives that such
provisions must support. This can be illustrated from the point of
view of a developing country such as India which, along with
countries in a similar position, identifies four types of provisions:
fundamental (no scope for negotiations); protecting (to defend and
legitimize developing country objectives); nurturing (to enhance
export and domestic capability); and bargaining (can be offered
selectively in order to achieve all or some of the first three provisions)
(Modivel, 1989). The most interesting of these provisions (in the
context of this volume) is the nurturing provision, which clearly
requires each country to look systematically at its service sector
activities to establish those that will benefit from liberalization and/
or supportive national measures. Modivel (1989) lists eleven service
activities in India that may gain from nurturing provisions. For
example, air transportation, for both domestic and international
services, is finding it difficult to cope but could be made more
efficient and offer better quality if domestic routes are opened to
foreign competition on the basis of reciprocating. Banking and
financial services will benefit from fuller use of information

technology and integration with global networks in order to develop export potential. Project and consultancy services have a strong export potential but stronger supportive measures are required. Finally, computer software is already a very high-growth sector supported by government policies but Indian nationals face problems gaining access and entry permits to allow them to deliver these services abroad, and improved access to information networks and distribution channels would also help. Overall the service activities in India that stand to gain most from nurturing provisions are: civil construction project exports involving movement of equipment and human resources (skilled and unskilled), consultancy, operation and maintenance contracts for airports, power stations etc., computer software services, shipping, and some financial services (largely in the banking field) (Modivel, 1989; see also Mody, 1989).

By mid 1990 the US and the European Community had produced a draft text for a General Agreement on Trade in Services (GATS). This calls on countries to liberalize trade in services as fully as possible, apart from any specific reservations they may include on their national schedules. Provisions such as non-discrimination, transparency, cross-border sales, the movement of persons to sell services, rights of establishment licensing and certification should all be permitted unless included in any country's specific reservations. This approval is likely to be much too comprehensive and rapid for the developing countries, who favour a more global approach to eliminating obstacles to free trade. The US is not concerned, as is the EC, that GATS should cover all services; for example, it wishes to exclude or to negotiate separate agreements for telecommunications and shipping. Some key players of global trade in services remain very reluctant to deregulate some of their service markets. Japan has resolutely moved very slowly in the deregulation of interest rates, money markets and foreign exchange controls and in allowing foreign concerns to act as lead managers and underwriters on corporate issues and the import of financial products to Japan. The US, in particular, is pressing for greater reciprocity since Japanese service companies, especially in finance and banking, are able to operate on a more equal footing alongside their competitors in the US financial centres. According to Bhagwatti (1991), however, the closed character of the Japanese market is a myth; people believe that it is closed because they are told that it is by lobbyists in the US and elsewhere. Other countries are more prepared to introduce progressive changes. South Korea, for example, indicated early in 1991 that it will allow more foreign access to its service markets by gradually opening up the construction, transportation and telecommunications fields (*Financial Times*, 1991b). Most of the concessions will come into force in the second half of the 1990s when the results of the current GATT negotiations come to fruition. Foreigners will be permitted to hold up to 99 per cent of the shares in joint ventures with Korean international passenger and freight shipping, and branches and subsidiaries of foreign advertising companies have been permitted from 1991.

FOUR

Services and the Global System of Cities

Large cities are the gatekeepers of the world service economy. Much the same applies within national urban systems where 'the shift to the services has transformed the system of cities in fundamental ways' (Noyelle, 1985, 42; see also Knight, 1989). The transition to a post-industrial world, the increased diversity of international trade to include services, and the rise of the service MNE have helped to consolidate the pivotal role of cities. Particular significance has been attached to the increasing mobility of capital (Castells, 1989; Bina and Yaghmaian, 1991) as a result of, firstly, the mismatch between capital accumulation and investment opportunities in a number of advanced economies and, secondly, the restructuring of financial institutions and markets. We will return to restructuring later in this chapter.

Whether one is supplying or consuming a service, it is necessary to make decisions about where to locate the individuals or the establishments that mediate these processes. Such decisions are important within local or national space economies; they become very important when they involve transactions in the global marketplace. Trading in international markets introduces even greater uncertainty because of factors such as cultural differences, local regulatory requirements and the time necessary to compile a portfolio of local clients and (if necessary) suppliers. Careful and conservative location decisions, perhaps modelled on the behaviour of competitors, is one way to minimize the uncertainty of expanding into new environments. Thus among the most notable attributes of the geographical expansion of services at international level have been concentration and centralization, especially in large cities and often at the expense of smaller cities.

There are a number of general factors that steer service firms operating at the international level towards urban areas. Reference has already been made to the minimization of uncertainty; this refers to the risks involved in committing personnel and financial resources to markets with different cultural, political, social or regulatory environments. The instinct of many service firms is therefore to make sure that they locate in those places that will give them access to as much information as possible on market needs, client expectations, and business practices and regulations. This may involve contact with local businesses but, crucially, may also allow them to monitor the behaviour of competitors whose location decisions are often influential. Shadowing competitors may seem an unlikely way to capture a share of new markets but, in some services such as banking or finance, pioneering location choices are not made lightly because of the adverse signals that they may be perceived to convey to investors or businesses. It is clearly difficult to quantify the effect of this follow-the-leader style of locational decision making, but it does appear to be important even if service firms will tend to rationalize their decisions by referring to the more conventional influences on location decision making.

The best known of these conventional influences are urbanization economies and localization economies. The former refer to the benefits accruing to services (and other producers) from access to the infrastructure of transport, telecommunications, housing, office buildings, and large and diverse labour markets found in urban areas. The latter refer to the benefits of proximity to similar services or to other economic activities that provide inputs to, or are clients of, service enterprises. The latter are especially important in the context of reducing uncertainty in location decision making. Another factor that has become crucial to effective location decision making is the kind and quality of labour – the human resource factor. As a proportion of their total labour inputs (and costs), services in general require more managerial, professional and technical workers. Many of these must be trained to high levels of specialization and will have first, and sometimes higher, degrees from universities or similar institutions (Bertrand and Noyelle, 1986; Guile and Quinn, 1988; Bannon and Tarbatt, 1990).

Flexibility in the labour force is also important because job specifications or the skills expected of individuals change rapidly in services such as computing or finance which are attempting to retain or expand markets through product innovation or the refinement of existing products.

There are obviously links between urbanization and localization economies and the diversity (although not necessarily the quality) of human resources. Each tends to reinforce the other with the assistance, at the global scale, of the pattern of international air and sea transport and telecommunications infrastructure and investment. It has also been stressed earlier that there are strong links between many services and manufacturing industry, so much so that as manufacturing multinationals emerged before service multinationals the latter have tended to follow the location decisions of the former. This applies particularly to the location of overseas corporate headquarters rather than to production plant location. Service firms supplying highly specialized, technical advice or management services connected with all aspects of international financial transactions are also influenced in their location decision making by the growing complexity of the arrangements involved: private and public sector organizations; funds in various currencies and from numerous different services; and projects involving numerous partners or subcontractors from different countries with different legal requirements for corporate operations. The location choices available to such firms (corporate legal services, management consultants, for example) are actually rather limited if they are to ensure that they have access to the best possible information, labour and specialist advice needed to fulfil their obligations to their clients.

Against this background, we can turn to some more specific explanations for the close relationship between the globalization of services and the evolution of the urban system. The transactional activities generated by services, mediated by telecommunications, and expressed in the spatial and market strategies of service firms are particularly important and are examinined in more detail below. These have also contributed to the development of so-called world cities (Friedmann and Wolff, 1982) or global cities (Sassen, 1992) which are organized into a clear, but changing, hierarchy. The emergence of these cities and the associated hierarchy will be discussed in the second part of this chapter; there follows an outline of some of the explanations for the association between the recent expansion of services and the form and structure of the global urban system.

Services and the Global Urban System: Some Explanations

Transactional forces

The symbiosis between the internationalization of service firms and concentration in world cities may be explained by considering the transaction costs of the former (Williamson, 1979). Since service firms are engaged in exchanges of information between supplier and client that require active participation by both parties, the transaction costs of service firms are usually higher than for manufacturing firms. We do not need to have the guidance of the manufacturer to drive a car which we have purchased, but a manufacturing firm will need to talk regularly with its corporate legal advisers if it is acquiring a company in another country (with different legal systems, methods of drawing up contracts or requirements relating to financing). Similarly, the purchase of computer software to handle the payroll of a large company will almost certainly need the on-the spot support of the supplier with installation, tailoring of the program to suit the needs of the client, ironing out any initial problems and installing any future upgrades. Complexity and reciprocity are more prominent in service than in manufacturing transactions. Investment in human capital ensures that the quality of a transaction is maintained but it increases the cost. Thus, as transaction costs increase, suppliers and clients move closer together and this leads to concentration at international level in major cities. But forward transaction costs with clients are only part of the explanation. Service firms also encounter backward transaction costs with other services that are sources of input to their outputs. These relationships are more diverse than those with clients (who may be a relatively small number of firms or individuals). The need to minimize transaction costs arising from the exchange of inputs therefore also promotes concentration in large cities.

The transactional approach to explaining world city formation and the contribution of services to the process is most strongly developed in the extensive writings of Gottman (1970; 1983; see also Corey, 1982). The transactional city is characterized by the structural dominance of white-collar employment; the domination of the city centre by knowledge processing, decision making functions typified by the control of multinational service corporations; the presence of intermediate services that support the control functions to create a corporate complex (Cohen, 1979); and the complementarity of local face-to-face transactions and long-distance telecommunications transactions. The latter ensure that the 'transactional metropolis is interwoven with other transactional centres, forming metropolitan networks within national territories and across international boundaries' (Corey, 1982, 421).

Information technology and telecommunications

The concept of the transactional metropolis or the network of world cities and the contribution of services to its development is very dependent on the mediating role of information technology and telecommunications. Flows of global capital and the rise of global cities would be impossible without it. When the IT revolution began to accelerate during the 1960s it was expected that it would release economic activities in general, and decision making functions in particular, from the restricted locational choices open to them using more conventional communications systems. IT could substitute for face-to-face communications since business could be transacted using electronic media. This would be especially significant for long-distance or intercity transactions which could now be undertaken between two parties located wherever they had access to telecommunications facilities. The outcome would be a breakdown of the traditional pull of established (or emerging) global service centres on the location of services moving into the international arena (similar effects would also occur at the national level as well as within individual cities). In other words IT and telecommunications would decentralize services within the urban system and within cities. On a purely intuitive basis there is some logic in this line of reasoning: service MNEs that are anxious to control labour costs can more effectively gain access to non-metropolitan, low-wage-labour markets; and high costs of congestion are imposed on residents and companies in large cities through time lost in delays, air pollution, high office rents and housing costs, and reduced quality of life.

In practice, however, there is a distinction to be made between the intent or ability to use information technology for the production of services, and access to the telecommunications facilities and networks that are required. Whether these are provided by publicly owned or private companies (provision varies in different countries), investment in the necessary infrastructure (copper, coaxial or fibre-optic cables, repeater stations, switching systems, channels, transmitters, satellite dishes etc.) tends to concentrate on the most likely sources of demand since the installation costs are high and recouping these costs is a vital priority, especially for privately owned telecommunications companies. The principal sources of demand are those companies and headquarters activities that require a wide range of information in order to be able to successfully coordinate their international activities. We have seen how these are largely isolated in the international information capitals (Moss, 1987) such as London, New York, Tokyo, Los Angeles, Hong Kong and Singapore.

Investment in telecommunications, especially any new investment in state-of-the-art facilities such as a teleport – a satellite-based global telecommunications network which gives users access to communication satellites considered vital for high-speed (and relatively uncongested) transmission of voice, data and video signals. Even more important for yet faster long-distance transmission of information is fibre-optic cable, which has a much higher capacity than copper cable and is also much more secure. Again, however, the economic necessities of fibre-optic technology require city-to-city links (or links between international information centres). Within countries, extensive use can be made of land adjacent to railway or major highway routes between cities, while between countries, especially across the Atlantic and the Pacific, a small number of trunk cables accommodate the information exchange requirements of users.

There is therefore a close association between the evolving geography of global telecommunications services and the development of cities that are nodes in the system. Whatever form these networks take (Langdale, 1989 generalizes them into three types:

global, regional hub-and-spoke, centralized) it is essential for services attempting to enter the international markets or attempting to improve coordination of their existing transnational facilities to be located in or very near to nodes in the global telecommunication network. With most large TNCs having headquarters (international, regional or national) in the international information capitals, it is not surprising that they also account for a significant proportion of international business telecommunications traffic. Many of them use not only public telecommunications services but also private international leased networks (Langdale, 1985; 1989; 1991). This reflects the overwhelming importance of intracorporate communication for international business transactions and also explains why investment in the facilities they require will tend to be concentrated in the cities where the outgoing and incoming information originates.

An example of a city that has benefited from this process and ensured its secure position in the global urban system is New York (Moss, 1986). Supported by deregulation of the telecommunications industry, New York has a more diverse and extensive telecommunications infrastructure than any other city in the world: one local business leader states that Manhattan has more than 'twice the telecommunications switching capacity of the average foreign country, more computers than a country the size of Brazil, and more word processors than all the countries of Europe combined' (quoted in Moss, 1986, 383). Apart from its diverse intracity telecommunications systems, including expanding fibre-optic networks, cable television systems, terrestrial microwave systems, cellular mobile radio and 'smart' office buildings, New York's long-distance, national and international telecommunications facilities are of particular significance here. At the national level, US Telecom provides a 37,000 (23,000 mile) fibre-optic network linking New York with other major US cities, and BTE provides a regional fibre-optic network linking suburban centres with the region. For international communications the teleport (serving seventeen earth stations) on Staten Island provides advanced data and voice services around the clock for users in New York and New Jersey. Thus, in New York major teleport service nodes include the World Trade Center, the Empire State Building and 77 Wall Street (in the heart of the financial area of Manhattan). There are also several companies operating communication satellite facilities providing access to earth stations on a business carrier basis. These include ITT, whose Manhattan earth station will provide ITT customers with intracorporate or client connections in more than 200 countries around the world. In addition to these satellite common carriers there are also a number of public communication networks such as GTE Telenet (links to more than 250 cities and more than 40 countries) or Tymnet (links to 500 US cities and, via connections with international second carriers, to more than 50 countries).

Moss (1987; 1991) has also produced some empirical evidence to support the contention that the strength of the New York telecommunications infrastructure and the international links that it facilitates are reflected in its share of information transactions. The spin-off does not necessarily require information exchange using telecommunications. Indeed, it is difficult to obtain data on telecommunications-based information flows because of the deregulation and privatization in recent years. Thus, more than one in five of the 325,469 documents and packages delivered from Japan to the US by DHL Worldwide Ltd in 1985 were destined for New York City. Similarly, New York City, San Francisco and Los Angeles are the origin of 45 per cent of all the documents transported by DHL to Japan (Moss, 1987). Although the figure is lower than in 1982, almost 19 per cent of the air passengers using New York airports in 1987 (for 21 city pairs with 100,000 or more passengers) were arriving from or leaving for London (Moss, 1991).

Spatial and market decision making

The spatial and market decision making of service firms is also central to any explanation of the evolution of the national and global system of cities. Camagni (1990) suggests that there are three kinds of spatial behaviour that have evolved through time and which can be used to interpret change in the structure of the urban system. The most widespread and long established is territorial behaviour: firms buy and sell within areas (markets) with geographically definable limits. Transport costs determine the limits to these markets (retail services, for example) and result in a distribution of cities according to the principles proposed by Christaller (1933). In the second case, where competitive behaviour is uppermost, service firms are not confined to a delimited geographical market area but compete to obtain a share (preferably as large as possible) of the national or global market. Marketing then becomes a key part of the production of a service rather than the cost of transport to the market. Services are exchanged between markets, and firms become organized into specialized units such as marketing, research and development, management services or personnel. Specialization allows firms to take advantage of economies of scale as well as of the localization economies mentioned earlier. The third and most recent form of spatial behaviour is based on the organization of networks (Batty, 1991). This is a response to the need for firms to be innovative in order to stay ahead of the competition; this requires state-of-the-art knowledge in highly specialized activities, and the easiest way to acquire the information and expertise is to establish cooperative and strategic arrangements with other firms. Scale economies are achieved through cooperation and the sharing of the advantages amongst the participants in a network. The resulting networks are constrained not by geographical limits to markets but by the availability of suitable media such as telecommunications for exchanging information quickly and, where necessary, confidentially. Nevertheless, networking requires firms to locate in the nodes that comprise the information technology and telecommunications infrastructure, the so-called 'wired cities' (Dutton, Blumler and Kraemer, 1987).

Neither competitive behaviour nor networking behaviour by service firms conforms well with the principles of central place theory, especially the latter. Because service firms of different size and function may have linkages with each other (and likewise cities) there is a shift of emphasis from vertical linkages between cities or between firms to horizontal linkages. Those parts of the urban system in which primary industries or manufacturing and population-related (consumer) services are dominant will be organized along Christallerian lines. But those parts of the urban system in which manufacturing and producer services are prominent components of urban economies will be organized (spatially and functionally) as specialized and complementary networks of cities or as networks of cooperating (similar) cities (figure 4.1). At the top is a network of world cities dependent on international telecommunication networks (as well as physical transport networks)

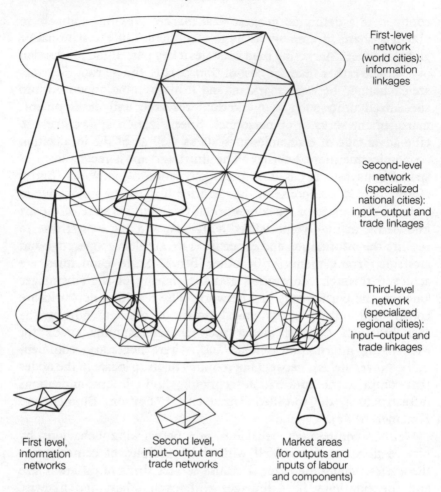

First-level
network
(world cities):
information
linkages

Second-level
network
(specialized
national cities):
input–output and
trade linkages

Third-level
network
(specialized
regional cities):
input–output and
trade linkages

First level,
information
networks

Second level,
input–output and
trade networks

Market areas
(for outputs and
inputs of labour
and components)

Figure 4.1   Hierarchy of city networks
Source: Camagni, 1990

for competition and cooperation. This top tier is the destination for
upward linkages from the second tier of specialized national cities
which are themselves linked, horizontally, by trade and other
activities. Downward, hierarchical relationships of the kind specified by Christaller will operate between the first and second tiers and
between the second and third tiers. The third tier comprises regional
specialized networks with horizontal linkages similar to those
operating at the second tier.

Services and World Cities in the Global Urban System

Identifying world cities

What then are the criteria that can be used to identify world cities and the hierarchy into which they are organized? The possibilities are of course numerous, but Friedmann (1986) lists the following: headquarters offices of transnational corporations (TNCs); regional headquarters offices of TNCs; rapid growth of business services; an important manufacturing centre; a major financial centre; a major transportation centre; and population size. Other criteria might include volume of international telephone traffic; rapid growth of office floorspace; diversity and quality of educational services; scale of research and development activities; range of cultural facilities; and high living and commercial business costs. Most of these criteria can be determined using data from a wide variety of published sources. Thus, the control function of world cities can be demonstrated by analysing the distribution of the headquarters of the 500 largest TNCs in the world in 1984 (Feagin and Smith, 1987). Ten cities account for just over 50 per cent of the top 500: New York (59), London (37), Tokyo (34), Paris (26), Chicago (18), Essen (18), Osaka (15), Los Angeles (14), Houston (11), Hamburg (10) and Dallas (10). Notice that New York has almost twice as many headquarters as London which, in turn, has rather more than Paris and Chicago.

Hierarchical principles can be demonstrated this way but need not necessarily operate consistently using different measures. Data showing the number of banks and traders engaged in foreign exchange (forex) trading, for example, reveal that London accommodated over twice as many as New York and ten times as many as Tokyo in 1985 (table 4.1) (Levich and Walter, 1989). The same table also demonstrates the concentration of these activities in a number of Western European cities, giving this region considerable comparative advantage over North America and Asia and the Middle East. Levich and Walter (1989) note, however, that while this interpretation seems reasonable, it may also represent a loss of efficiency in Western Europe because it operates with multiple currencies rather than as a more unified currency area such as North America. More recent statistics compiled in 1989 for eight different currency

Table 4.1   Location of banks and traders in foreign exchange, 1985

Location	No. of banks	No. of traders
Asia and Middle East		
Tokyo	30	169
Singapore	69	293
Hong Kong	52	306
Bahrain	38	176
Total	189	941
North America		
New York	108	793
Toronto	18	120
Chicago	14	81
Los Angeles	11	46
Total	158	1083
Western Europe		
London	258	1603
Luxembourg	74	353
Paris	75	459
Zurich	35	245
Frankfurt	48	334
Milan	37	191
Brussels	33	201
Total	560	3386

Source: extracted from Levich and Walter, 1989, 68, table 4.9

quotations on Reuters screens in fifteen leading financial centres show that just over 17 per cent of the total quotes were provided by London, followed by New York at 15 per cent, Singapore at 11.2 per cent and Hong Kong at 11 per cent. Applying differing measures to the same group of services activities also produces different ranks for cities around the world (table 4.2). Noyelle (1989) ranks the top twelve banking cities in relation to cumulated net income, cumulated assets and the average net income/asset ratio of the top 50 commercial banks and the top 25 securities firms in 1986. Tokyo, New York and London lead the ranking list by income, Paris displaces London in the assets ranking and Los Angeles displaces Tokyo for the income/asset ratio.

Table 4.2   Top twelve banking centres in the world,[a] 1986

| City | No. of firms | Ranked by:[b] | | |
		Income ($m)	Assets[c] ($b)	Net income/ asset ratio[d] (%)
Tokyo	22	6424 (1)	1801.4 (1)	0.357 (10)
New York	16	5673 (2)	904.8 (2)	0.627 (2)
London	5	2934 (3)	390.3 (5)	0.752 (1)
Paris	6	1712 (4)	659.3 (3)	0.260 (12)
Osaka	4	1261 (5)	557.6 (4)	–
Frankfurt	3	1003 (6)	306.8 (6)	0.327 (11)
Zurich	2	826 (7)	–	0.524 (4)
Amsterdam	3	739 (8)	193.4 (7)	0.382 (9)
Basel	1	415 (9)	–	0.489 (7)
Hong Kong	1	392 (10)	90.8 (12)	0.432 (8)
Los Angeles	1	386 (11)	–	0.617 (3)
Montreal	1	354 (12)	–	0.520 (6)

[a] Data for top 50 commercial banks, top 25 securities firms.
[b] Rank in brackets.
[c] Other cities are: Munich (8), Nagoya (9), San Francisco (10), Kobe (11).
[d] Other city: Toronto (4).
Source: extracted from Noyelle, 1989, 101, table 5.5

A univariate classification will produce different results depending upon the variables selected. Geneva, Brussels, Madrid and Vienna, for example, come much higher up the rankings on the basis of the number of international conventions (table 4.3). International financial centres in the first rank, notably New York and Tokyo, are not amongst the top ten international convention centres. An alternative ranking according to the number of headquarters and regional secretariats of international organizations also pushes New York and Tokyo down the list while London remains near the top (table 4.4). Cities in the less developed or the newly industrialized countries also rank among the top twenty centres – an indication of the role of international organizations as agents for the distribution of finance, expertise and other assistance in these areas.

In overall terms, however, it will be apparent that there is still considerable overlap of functions between the cities included in lists of this kind; the rankings may vary but the cities included are

Table 4.3   Number of international conventions, by city, 1982 and 1986

| City | 1982 | | 1986 | |
	Number of conventions	Rank	Number of conventions	Rank
Paris	292	1	358	1
London	242	2	258	2
Geneva	147	3	180	4
Brussels	118	4	157	4
Madrid	22	33	118	5
Vienna	90	5	106	6
W. Berlin	47	11	100	7
Singapore	44	14	100	8
Barcelona	8	–	96	9
Amsterdam	36	19	84	10
Seoul	36	19	84	11
Washington DC	45	13	75	12
New York	70	6	72	13
Rome	69	8	69	14
Strasbourg	52	10	67	15
Munich	34	23	63	16
Copenhagen	70	6	63	17
Stockholm	28	27	63	18
Tokyo	55	9	56	19
Hong Kong	46	12	54	20

Source: extracted from Knight and Gappert, 1989, table 22.1

generally much the same. The resulting world city hierarchy has been identified by Friedmann (1986) although he uses all seven of the criteria for world city status (see above) (table 4.5). There are many more primary centres in the core countries (developed market economies) than in the semi-periphery (upper-middle-income market economies) where many secondary cities rival the core primary cities on the basis of size but not many of the other criteria for world city status. If, as Friedmann (1986, 69) suggests, 'the economic variable is likely to be decisive for all attempts at explanation' of the world city and the complex spatial hierarchy shown in table 4.5, then it is clear that service industries are helping to underwrite existing and future relationships between cities (producer services) as well as the restructuring that is taking place within cities (see chapter 5). Indeed the level of integration between

Table 4.4   Location of international organizations[a] by city, 1987 and 1988

Rank	City	No. of international organizations	Rank	City	No. of international organizations
1	Paris	866	11	Strasbourg	93
2	Brussels	862	12	Zurich	89
3	London	493	13	Oslo	88
4	Rome	445	14	Bangkok	82
5	Geneva	397	15	Helsinki	75
6	New York	232	16	Nairobi	75
7	Washington DC	180	17	Mexico City	69
8	Stockholm	128	18	Caracas	68
9	Vienna	115	19	The Hague	67
10	Copenhagen	114	20	Tokyo	65

[a] Number of headquarters and regional secretariats.
Source: extracted from Knight and Gappert, 1989, table 1.2

Table 4.5   World city hierarchy[a]

Core countries		Semi-periphery countries	
Primary cities	Secondary cities	Primary cities	Secondary cities
London*	*Brussels***		
Paris*	*Milan*		
Rotterdam	*Vienna***		
Frankfurt	*Madrid***		
Zurich			*Johannesburg*
New York	*Toronto*	*São Paulo*	**Buenos Aires***
Chicago	*Miami*		**Rio de Janeiro**
Los Angeles	*Houston*		*Caracas***
	San Francisco		**Mexico City***
Tokyo*	*Sydney*	*Singapore***	Hong Kong
			*Taipei***
			Manila*
			Bangkok*
			Seoul*

[a] Cities classified by population of metropolitan region: **10–20 million**; 5–10 million; *1–5 million*.
* National capital.
Source: Friedmann, 1986, 72

a city and the world economy and the functions that it performs within the new international division of labour are crucial to the scale and characteristics of internal structural change (see chapter 5).

The emergence of world cities

Although there is obviously scope for argument about how best to identify world cities, there is no doubt that they play a dominant role in the operation of the global space economy. This dominance is increasingly articulated by the requirements of the world service economy. The internationalization of production, initially of goods but now including services; the internationalization of finance; and the internationalization of the state have played a crucial part in this process (Thrift, 1986). Theoretical and empirical studies of the world city phenomenon have become commonplace in recent years (Friedmann and Wolff, 1982; Friedmann, 1986). An early study which drew attention to the disproportionate concentration of service businesses in very large cities was undertaken by Hall (1966) who suggests that the concept has much earlier beginnings in the writings of Geddes (1915). Hall was clearly concerned to demonstrate the development and planning of individual world cities, whereas Friedmann and Wolff (1982) explore the relationship between an increasing integration of world production of goods and services and a global network of independent cities. As the world economy has become increasingly controlled by MNEs, for example, the role of world cities has steadily become stronger. The global economy in which services is playing a growing part requires 'control points' where corporate headquarters, financial intermediaries and highly specialized services group together, for the reasons outlined earlier, to convey decisions or to conduct transactions over large areas of the globe and in the shortest possible time.

The interaction between information circulation and the multiplier effects of the location decisions by service (and manufacturing) organizations gives rise to the process of cumulative causation (or reinforcement) (Pred, 1973; 1977; Borchert, 1978; Dunning and Norman, 1987). This involves the growth of activities as a result of the intermetropolitan interactions of service and manufacturing MNEs, which in turn leads to a tendency for innovations that generate further growth, which in turn is transmitted between those cities that host MNE intermetropolitan interactions. As a consequence, decision making functions accumulate to a disproportionate degree in a few large cities along with their high-level specialist staff and officials whose presence makes these cities attractive for any further rounds of locational decision making (Ross and Trachter, 1983).

With growth impulses more likely to be transmitted between cities at the top end of the global urban hierarchy, the prospects for downward diffusion of economic growth and social change to second- or third-tier cities are rather limited. The increasing involvement of service industries in the process seems likely to exaggerate further the differences between core cities and peripheral cities or between first- and third-tier cities. In many ways this possibility mirrors the anxieties expressed by developing countries about liberalizing trade in services and the likelihood that the benefits will be even more concentrated in the developed market economies and, in particular, the major cities that manage the global economy.

Thus, the global system of cities can be analysed as a set of nodes joined together by a series of linkages or flows. The nodes (or cities) that comprise the system vary in location, size, role and the level of linkage (or connectivity) with other cities in the system (see for example Borchert, Bourne and Sinclair, 1986; Timberlake, 1985). The linkages are measured in terms of flows of information, people, trade or finance. The relationship between the cities in the system, as well as the performance of individual cities, is determined by 'external' factors such as technological change or organizational decisions concerning the location of production or distribution. Any changes in one city as a result of technological change are transmitted via the network of linkages to other cities.

As a general rule, the sizes of the origin and destination determine the strength of the links between them so that the largest cities will dominate to give the hierarchical structure to the system. In other words the largest cities will have links with most of the other cities in the system, irrespective of distance, while the smaller cities will tend to have their strongest links with a larger city nearest to them. Urban systems are quite well integrated at the national scale and this means that cities have become specialized in particular functions. This is well illustrated by Noyelle and Stanback's (1984) work on the restructuring of the US urban system which led them to classify metropolitan areas into diversified service centres, specialized service centres, production centres and consumer-oriented centres. They identify 39 diversified service centres which accounted for 34 per cent of US population (1977). The concentration of decision making functions was very marked in these centres: they were the location for 62 per cent of the national headquarters of the 1200 largest firms in the US, 67 per cent of the deposits held by the largest 250 commercial banks, and 72 per cent of the partners working in the eight largest accountancy practices. They also housed concentrations of distribution-related services: 83 per cent of the headquarters of the top 200 advertising firms, and 72 per cent of the regional headquarters of a sample of large corporations. Finally, the 39 cities accommodated a large share of technology-related functions: 52 per cent of the medical schools in the US, 41 per cent of the top 100 universities, 40 per cent of all R&D laboratories, and 52 per cent of the divisional head offices of the largest 650 firms in the US. These diversified service centres serve national markets (New York, Los Angeles, Chicago, San Francisco), regional markets (e.g. Philadelphia, Dallas) and sub regional markets (e.g. Memphis, Syracuse).

If national urban systems are being transformed in this way by services and we know that the global economic system is becoming more open and interactive, then we might expect the global system of cities to be subject to similar processes. It is less easy to illustrate this with data of the kind used for the US urban system, but it is possible to piece together recent theoretical work and empirical evidence about the impact of services on the global urban system (Rimmer, 1991; Daly, 1991). We have already seen how the expansion of service MNEs has introduced greater specialization and integration into international trade; the large institutions and corporate organizations involved can be expected to continue growing and to exert even more influence on the urban system through their location decisions and investment priorities if only because most of these decisions are already taken in the cities that are established nodes in the national and international urban system.

Services and the Global Urban System: Some Examples

International financial centres

Developments in the global financial system have been responsible for a dramatic escalation in the volume of international financial transactions during the 1980s (Hamilton, 1986; Mendelsohn, 1980; Cooper et al., 1987; Jones, 1992). It is useful to outline some of the key features. There are at least three principal reasons for these trends: deregulation of domestic financial markets and controls, innovation in the services available from financial markets, and advances in technology (Leyshon, Thrift and Daniels, 1987a; Davis and Latter, 1989; Dicken, 1992). It has been suggested that regulation of domestic (national) financial markets induces structural weaknesses that actually make the constituent services vulnerable to international competition (Price, 1986). Even minor regulatory changes will therefore encourage finance capital to move around the system; only if national finance markets are all equally regulated will world capital remain immobile. A frequently cited example of regulation encouraging international monetary flows is the development of Euromarkets in which Eurocurrencies (such as the Eurodollar or Euroyen) are traded as financial commodities exchangeable outside their country of issue. These markets arise from the differentials in regulatory control in the US and European financial markets. The Euromarkets were originally based in London and were actively encouraged by the Bank of England on the ground, amongst others, that it would consolidate London's status as an emerging international financial centre. Another change that has encouraged institutional investors in domestic markets to move more of their funds offshore has been the removal of exchange controls on capital flowing from internal to external markets. This has triggered an ongoing process of domestic financial system deregulation in countries across the world as each tries to create an attractive environment for international financial institutions. However, it has been argued that the deregulation encouraged by the competitive struggle between the global financial centres is an illusion, i.e. state regulation of rule systems formally devised by the companies that use them is the reality, which is quite the opposite of common perceptions of deregulation (see Moran, 1991). In each case the effects of deregulation are most strongly felt by the leading city or cities where the majority of financial activities are concentrated, especially their interaction with other, similar cities in other countries or global regions.

Innovation in the instruments used to facilitate international debt has created a bewildering array of options in recent years (see for example Leyshon, Thrift and Daniels, 1987a). During the 1970s debt was largely arranged in the form of loans, principally through the Euromarkets. Such Euroloans were organized through syndicates of banks and these therefore required a presence in the leading Euromarket centres. London was a key city in the market: and the number of foreign banks represented there almost tripled between 1968 and 1986 (from 134 to 447) and the associated employment increased by 500 per cent (Blanden, 1986). More than 60 per cent of the syndicated loans analysed in 1979 emanated from London (Plender and Wallace, 1985). But loans of any kind are only profitable if the debtor can repay with interest; many of the developing countries, in particular, found it difficult to honour their commitments and it is estimated that by 1984 some 35 sovereign borrowers were unable to service their debt. A crisis developed which threatened the world financial system and ultimately resulted in an emphasis on securitization of debt in favour of syndication. A foreign bond or a Eurobond is a security issued by a borrower giving the holder a specified income at an agreed maturity date and interests payments at regular intervals before that date. Eurobonds may be available in several Eurocurrencies and may be sold in several countries at the same time. Foreign bonds are issued in a country other than that of the issuer and are usually purchased by that country's investors. The Eurobond market was valued at only $147m in 1963, growing to almost $195b in 1986 (OECD, 1986). It is much larger than the foreign bond market, where issues were valued at $35b in 1986.

Banks manage the issue of new bonds and also trade in bonds that have already been issued (the so-called secondary market). Since the bond markets can only work if investors purchase them (thus exposing far less of a bank's capital to the lending process) it is important to have highly effective methods of distributing them. The faster the distribution system, the shorter the exposure of the bank; a wider distribution network will also affect this consideration. International distribution networks, especially linking other sources of bond issues, are therefore vital. The market for new and secondary trading is centred on London, with strong links to Tokyo, New York and a number of secondary financial centres. Banks and securities houses from Japan, Western Europe and North America dominate the market. Thus the value of bonds issued in the first half of 1986 was $57b; almost 39 per cent of issues were handled by just five leading managers, i.e. Credit Suisse First Boston (US/ Switzerland), Deutsche Bank (Germany), Merrill Lynch (US), Salomon Bros (US) and Nomura Securities (Japan). There are also numerous variants on the fixed-rate bonds together with other methods of securing debt obligations such as currency or interest rate swaps or 'junk' bonds. Competition is fierce, the staffing expertise required is considerable and the market is largely international. Large cities with the requisite infrastructure of telecommunications, suitable office space, a pool of supporting specialist services and labour and a regulatory environment conducive to bond operations are inevitably the key locations for these activities. This has become more so as the shift from syndicated lending to securitized lending has continued; bonds can be traded between locations and clients in a way that was not possible with loan debt. Thus, many securities are traded globally on a continuous basis using very sophisticated telecommunications networks connecting the issuing houses and traders in cities in different time zones around the globe. A triad of cities is the key to the system. When the London trading day closes, firms can pass on their books to New York (which is still open) and, when it closes, firms can pass on their books to their Tokyo partners and offices, which will just be closing when the market opens again in London (see fig. 2.2). In this way firms can respond very quickly to market trends, especially sudden movements, and can operate with less risk and on smaller margins (Lambert, 1986). The latter may give a competitive edge.

A further stage in the globalization of capital markets is the buying and selling of shares (equities) in private firms. The values of Euro-equity issues increased by a factor of seven from $1200m to $7900m, between 1984 and 1986 alone. MNEs can stabilize their operational bases by diversifying their shareholdings. This may also enhance their status as truly global corporations. Pension funds that invest in a variety of financial instruments are attracted by Euro-equities as a way of diversifying their portfolios. Although the number of firms offering shares in this way remains modest it is expected to grow during the 1990s. The leading stock markets for foreign listings are London (505), Amsterdam (278), Zurich (176) and Frankfurt (189) (*Euromoney*, 1986).

Finally, world currency markets operate in an environment of floating exchange rates. Uncertainty about currency values from one day to the next is endemic and has encouraged financial institutions to find ways of minimizing its effects. The result has been the creation of markets in financial futures whereby 'contracts are made for the purchase or sale of standardized bundles of financial instruments at agreed prices to expire at set periodic intervals' (Leyshon, Thrift and Daniels, 1987a). As with bonds, futures and tradables have therefore added to further securitization of finance. They have also given rise to specialist exchanges following the lead given by the Chicago Mercantile Exchange where the first financial futures were traded in 1972.

All this diversification and increasingly international orientation of financial services could not take place without the support of appropriate mechanisms to make available the information needed by participants, to allow dealing to take place, and to ensure prompt confirmation and settlement of deals. Information technology, already outlined in chapter 2, has made all this possible while also spawning services derived from the technology. As Hepworth (1989, 180–1) remarks, the 'most dramatic and visible impacts of the information economy have occurred in the world's capital markets' and 'information technology has become crucial . . . to the investment and locational strategies of the market's dominant users – multinational corporations.'

International financial centres (IFCs) have therefore evolved in parallel with the diversification of financial markets (Kindleberger, 1974; Reed, 1981; Coakley and Harris, 1983; Coakley, 1992). An IFC 'acts as a bank for the entire world and the services it renders facilitate the flow of goods, and services, and capital among nations' (Nader et al., 1955, quoted by Reed, 1981, 4) or, according to Kindleberger (1974, 57), performs 'the highly specialized functions of lending abroad and serving as a clearing house for payments among countries'. An IFC provides the full range of financial services, with a strong orientation to international transactions, as distinct from an international banking centre (IBC) which largely provides comprehensive banking services (Reed, 1981). Alternatively, the international links that can best be used to identify an IFC can be classified in three ways (Coakley, 1992). The first is if the client base or the sphere of influence of the financial services in a city extends beyond national boundaries. The second is derived from the extent to which the financial services that are headquartered in another country influence the supply and demand for financial services within a centre. The proportion of turnover transacted in foreign rather than domestic (host country) currency is the third dimension that can be used to classify an IFC.

IBCs are more numerous around the globe and may ultimately evolve into IFCs. Examples of IBCs are Frankfurt and Paris in Europe, San Francisco and Toronto in North America, Hong Kong and Singapore in Asia, and Sydney in Australasia. These cities are placed lower down in a hierarchy of global financial centres and, as Coakley (1992) suggests, might be better described as regional financial centres. Position in the hierarchy is therefore determined by the relationship between the centre and its market: the wider the geographical extent of the market, the nearer to the top of the hierarchy a centre becomes. Financial centres largely channelling international (and other financial services) to and from national markets will occupy a lower position than a centre that provides financial services to several contiguous countries or offers equities from firms in those countries on the stock market.

It is certainly the case that the existing IFCs were initially IBCs since a banking infrastructure is a prerequisite for the functioning of other financial services while also conferring power and status on the cities where the banks congregate. The principal IFCs, sometimes referred to as the Golden Triangle, are London, New York and Tokyo. But each has different strengths so that it is difficult to identify a pre-eminent centre at the top of the global hierarchy. London is the principal centre for dealing in foreign exchange but Tokyo is the leading centre for international banking. Indeed, none of these three centres is the leading centre for trading in futures and options; this position is occupied by Chicago which has the three leading international financial futures exchanges: the Chicago Board Options Exchange, the Chicago Board of Trade and the Mercantile

Exchange. The participants in financial futures markets are a good cross-section of the financial community – banks, discount houses, securities houses, stockbrokers and money brokers – so that other major exchanges (by volume of contracts) are also located in the principal IFCs, in particular New York and London (London International Financial Futures Exchange). Smaller futures exchanges have been established in a number of IBCs such as Toronto and Singapore.

The competition between cities participating in some or most aspects of the circulation of global finance capital ensures that the balance of 'power' within the global urban system is open to modification. This is demonstrated by the way in which London's share of international lending by banks decreased from 27.1 per cent in 1975 to 20.5 per cent in 1989, while for the same period Tokyo's share rose from 4.6 per cent to 20.6 per cent (Coakley, 1992; see also Coopers and Lybrand Deloitte, 1990). Although London had the largest foreign exchange turnover in 1986 and 1989 compared with New York and Tokyo, the change in the volume of trading was highest in Tokyo (140 per cent) followed by New York (120 per cent) and London (108 per cent). Stock exchange turnover is led by Tokyo, followed by New York and London; however, the position of London is increasingly threatened by Frankfurt, where the reunification of West and East Germany has raised its share of global stock market turnover ahead of London's.

Much of the speculation is connected with the advent of the Single European Market which might cause a shift in the European financial centre of gravity. It is also feared that the proposed European Community central bank, for which Frankfurt has been mooted as a location, will draw large amounts of business away from other European financial centres. In technical terms London's advantage is considerable: it has enormous breadth and depth of expertise and experience. The decision by the European Bank of Reconstruction and Development to choose London for its headquarters confirms this, as well as the UK's decision to enter into the European Monetary Union (EMU) at the appropriate time. This may now be several years away after the UK withdrew from the European exchange rate mechanism (ERM) in September 1992. Further uncertainty about London's dominant position as a financial centre will be caused by the virtual certainty that the

Table 4.6   Cost of living index and annualized inflation rate,
selected European cities, 1991

City	Living cost index	Inflation (%)
Helsinki	122.7	6.2
Stockholm	121.7	11.1
Geneva	118.4	6.1
Zurich	117.9	6.1
Copenhagen	112.7	2.1
Milan	107.8	6.3
Madrid	105.7	6.5
Rome	103.6	6.3
Barcelona	103.6	6.5
Paris	103.5	3.5
Vienna	103.1	3.2
Brussels	102.6	3.7
Frankfurt	100.7	2.9
Dublin	100.2	3.5
London	100	10.9
Moscow	−98.7	12.0
Lyon	−94.6	3.5
Amsterdam	−93.9	2.4
The Hague	−93.2	2.4
Hamburg	−93.1	2.9
Bonn	−88.8	2.9

Source: Financial Times, 1991[a]

European central bank will be located in Bonn. But there are other location factors that can influence corporate decision makers (London Planning Advisory Committee (LPAC), 1991) including infrastructure, quality of life, cost of living and training of the labour force. The LPAC report demonstrates that London does not compare well with Paris or Frankfurt on quality of infrastructure, and both are less congested. On the other hand, London compares very well with some leading European cities in relation to a cost of living index (including outlays on housing) (table 4.6). The result may be some erosion of London's pre-eminence as Europe's financial services centre unless more positive action is taken to protect its advantage by improving some of the shortcomings, especially in its transport infrastructure and training facilities (LPAC, 1991).

The relatively volatile status of IFCs and IBCs has persuaded some national or state governments to attempt some constructive intervention designed to influence the location decisions of international financial services. One example is Canada where an effort has been made to counter the flight of capital and the concentration of national and foreign financial and related services in Toronto. Since 1986 tax incentives have been offered by British Columbia, Quebec and to a limited extent the federal government in an effort to attract a growing number of foreign banks and asset management companies to Vancouver and Montreal (*The Times*, 1991a). By early 1991 a total of 47 Canadian and foreign institutions had registered (at $5000 each) as international financial centres (IFCs) in Vancouver and 27 had done so in Montreal. UK merchant banks S. G. Warburg and Schroders have consequently been attracted to Montreal, and an Edinburgh-based financial services firm decided to move its North American head office from Albany, New York to the city because of the IFC incentives. Other cities, such as London, have commissioned major studies that identify their strengths and weaknesses as international service centres. consider lessons from the past and outline the prospects for the year 2000 and beyond (LPAC, 1991). Comparison with the characteristics and behaviour of competitor global cities is always a major part of these studies.

In a more objective attempt to explain the emergence of IFCs, Goldberg, Helsley and Levi (1987; 1991) have used a number of international and national economic variables. Taking the size of the foreign assets of domestic banks and the proportion of total national employment in the financial sector as dependent variables, they use regression analyses to examine the effects of variables that can be grouped into four factors: the level of overall economic development; the importance of international trade; the extent of financial intermediation; and the stringency of financial regulation. The results are to some extent inconclusive in that no one factor consistently emerges as a significant explanation for the emergence of a country (in practice a major city) as an IFC. Imports most clearly emerge as positively associated with financial sector development; exports have the opposite effect. Nevertheless, foreign trade is found to be a positive determinant of financial sector growth, together with the regulatory environment. Tax incentives are positively related to the two dependent variables, while reserve requirements (the capital that must be held by banks to hedge against defaults on loans, for example) were negatively related.

Time series estimation was also performed on data for the UK, Canada and the US to see whether the results of the cross-sectional analysis were supported. Using the proportion of the labour force in the financial sector as the dependent variable, it is found that the results are in line with those produced by the cross-sectional analysis. Overall, the results confirm that the size of a national financial services sector, in particular banking services, is positively (and significantly) associated with the value of imports. Goldberg, Helsley and Levi (1987) suggest that this should not come as a surprise; the merchant banks of the UK or of the Netherlands, for example, developed precisely in order to finance import activities by other businesses and most of these have since become investment banks. Countries with high levels of export activity do not, as might be expected, have a high level of financial services activity; this may be because exporters borrow in the markets where they sell their goods and services in order to incur interest charges that will be payable in the same currency that is being used to purchase their product or service. Finally, the level of real GDP per capita is a positive influence on the size of the banking sector. As real GDP per capita rises the financial sector becomes larger. This leads Goldberg, Helsley and Levi (1987, 21–2) to conclude that 'international trade variables are responsible for most of the observed variation on the sizes of the financial sectors' in different countries at any one time or through time.

Other producer services

Financial services are not the only services that have internationalized strongly during the 1980s (UNCTC, 1990). Services such as accountancy and management consultancy firms (Leyshon, Thrift and Daniels, 1987a; Thrift, 1984; Moss, 1987), property consultants (Leyshon, Thrift and Daniels, 1987b), law firms (Moss, 1987), merchant banks and advertising (Perry, 1990) have also been active. The location patterns associated with this expansion are focused on a relatively small number of cities (figure 4.2); furthermore certain cities such as London, New York and Tokyo again feature

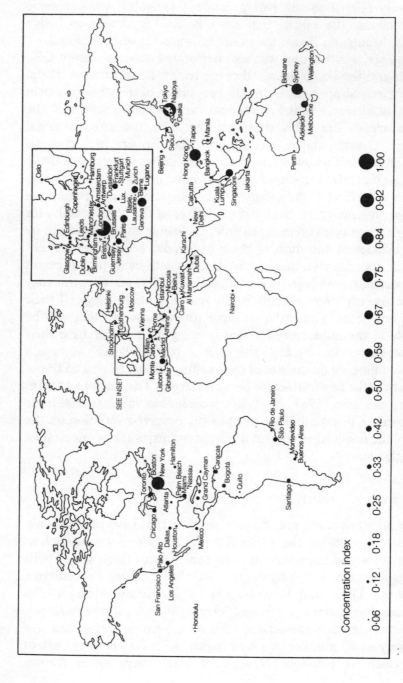

Figure 4.2 Head office, subsidiary, branch and representative offices of seventeen of the world's leading merchant and investment banks, 1985–1986

prominently in the location choices made by almost all the major service firms engaged in internationalization, while second-tier cities such as Sydney or Hong Kong that serve smaller regional markets have attracted a more limited cross-section. US legal firms, for example, have been following their clients to locations outside the US: in 1985, 48 of the top 500 firms had 72 offices in 11 European cities (notably London with 35 and Paris with 20) and 19 firms had 33 offices in 10 Pacific Rim cities (Hong Kong 13, Singapore 6) (Moss, 1987). The same cities were prominent locations for the offices of US advertising agencies in 1985–6 (London 30, Paris 25, Singapore 15, Hong Kong 17). Moss (1987, 542) comments that in Europe and the Pacific Rim 'there is a remarkably even distribution of offices between the largest information hubs of nation states.' Much the same can be said for the top international accounting firms (London 19, Paris 21). The hierarchical structure evident for financial service locations is replicated by other producer services, thus reinforcing the global control function of a limited number of cities supported by secondary cities performing regional control functions.

The competition between cities for the attraction of business services will also continue to intensify. Alessandrini, Secchi and Saviolo (1991) cite a number of recent attempts by cities to enhance their competitiveness as service locations. Lyon has embarked on Lyon 2010 which is designed to coordinate all the initiatives that will develop and promote the city as an international service centre (as well as its role as France's second city). Munich is investing in infrastructure that will support business location, including an airport project, while Barcelona has introduced policies that will develop the quality of public and private sector services as part of an economic and strategic plan for the year 2000. Milan is also increasing its already strong orientation towards a service economy by investing in a marketing plan (Alessandrini, Secchi and Saviolo, 1991).

The success of initiatives like these will be assisted by the diffusion of some services through the urban system. Evidence for this is most readily available at the scale of national urban systems. Thus, in the US there has been a demonstrable locational shift of producer services away from the nation's Standard Metropolitan Statistical Areas (SMSAs) throughout the 1970s and 1980s (Kirn,

1987; Noyelle, 1985; Moss and Dunau, 1986; Moss and Brion, 1988; see Howells, 1988 for an analysis of central and peripheral regions in Europe). Some of these losses are explained by companies shifting from the diversified and specialized service centres such as Chicago or New York to regional diversified service centres such as Houston or Atlanta; but the majority have relocated short distances to neighbouring SMSAs (see also Stanback and Noyelle, 1982). Overall, however, the numbers involved are small and services concerned with decision making have tended to remain in the largest diversified and specialized service centres (approximately 60 cities). Diffusion tends to take place amongst cities within the urban system that are of similar status. Hence, production and consumer service-oriented cities do not benefit. There is every reason for believing that this trend will continue and, if anything, will be reinforced by the shift towards the control of large parts of service production by large corporations. When this happens the smaller, more dispersed operations of the service firms absorbed will be centralized to create larger units.

The purpose of this chapter has been to illustrate the way in which the growing involvement of services in the world economy is reflected in the form and structure of the global urban system. The emphasis has been on interurban effects and relationships although, in practice, it is difficult to separate these from the economic and other circumstances within the cities that have attracted a large part of the service-related growth. We can therefore expect that the processes outlined in this chapter and in chapters 2 and 3 will bring about changes within cities as well as between them. In the next chapter we therefore turn to the effects of the internationalization of services on intra-urban restructuring.

FIVE

Internationalization of Services and Restructuring of Cities

Cities have always been important arenas for changes that reflect political, social or economic processes in the world around them. As the long-established foci for fulfilling the population- and production-based demand for services, cities have been able to offer superior transaction costs over those possible in rural areas. On the supply side, these lower transaction costs encourage entrepreneurial activity and the birth of new service enterprises that can draw upon the pool of expert knowledge, information and labour in the city. For some cities the transformation of the service sector during and since the 1970s ensured their economic renaissance. The best example is New York, which by 1976 was considered by many observers to be in the final throes of an inexorable decline (Warf, 1987). During the previous decade there had been a steady out-movement of population and of corporate headquarters; the property market had collapsed; the city's infrastructure was in a very poor state of repair; per capita incomes were falling; and, because many of these circumstances created losses of tax revenue, the New York City government was almost bankrupt. Since that crisis New York City and the surrounding metropolitan region has 'witnessed a remarkable resurgence. Led by healthy growth rates in financial and business services, New York in the 1980s has been transformed into one of the healthiest labour markets in the nation . . . its real estate prices and local government revenues have boomed' (Warf, 1987, 2; see also Drennan, 1991; Netzer, 1992). Although confronted by less fundamental crises than New York, many other cities have been transformed by the international and national expansion of services in the 1980s: London, Frankfurt, Los Angeles, Sydney, Tokyo, Singapore and Hong Kong are just a few examples.

In this chapter some of the impacts of the expansion of the service economy within cities are considered. The list of impacts is long and it will be necessary for some examples to suffice at the expense of conveying fully the range and depth of the impacts involved (Hall, 1991b). We can expect significant impacts on the employment structure of cities, and some comparative data for a number of large metropolitan areas around the world are discussed later in this chapter. There will also be local labour market effects, including an increase in the requirement for well educated, professional and specialized labour as well as in the demand for less skilled labour to provide 'back office' support for data entry, record keeping, invoicing and a wide range of related functions. Connected with this duopoly or polarization of labour demand is a widening gap between the incomes of the more and less skilled workers. This polarization may, in turn, lead to distinctions in residential location choice, in the requirements for educational and social services and in the range and types of retail services.

Other multiplier effects connected with the location choices of front and back office functions within cities will also emerge, especially with respect to the demand for producer services such as advertising, personnel recruitment, promotional services, computing and high-speed, high-quality printing (Daniels, 1987). The events of the last twenty years have also changed the way in which services are delivered in cities so that there will be consequences for physical planning, including the provision of transport infrastructures, the scale and layout of office and shopping complexes as personal mobility has improved, and the requirement for inner urban or suburban office development that will enable some services to escape the high costs (amongst other things) of locating in the central business district (CBD) (Brotchie et al. 1991). The characteristics of the buildings required by many service enterprises have also changed as economies of scale and the application of information technology have influenced the production and competitiveness of services (Daniels and Bobe, 1990).

Services and Employment Restructuring in Large Metropolitan Areas: Some Comparisons

Before looking more closely at some of the impacts of the growth of services on cities, it may be helpful to examine the evidence for the shifts that have triggered these effects. Broadly comparable statistics for a selection of cities are provided in table 5.1 (Daniels, 1990; see also Daniels, Hutton and O'Connor, 1992). New York and Tokyo

Table 5.1   Restructuring of metropolitan economies, selected cities, 1977–1987

| | Employment (000s) | | |
Year and change	Manufacturing	Services	Total
New York			
1977	1347	4731	6336[a]
1987	1127	5973	7532[a]
Change 1977–87	−220	1242	1196
Change (%)	−16.4	26.2	19.0
Île-de-France (Paris)			
1975	1277	1851	3128
1985	991	2094	3085
Change 1975–85	−286	243	−43
Change (%)	−22.4	13.2	−1.4
Melbourne			
1976	315	73	1263[b]
1986	259	843	1263[b]
Change 1976–86	−5	109	106
Change (%)	−17.6	14.9	9.0
Tokyo			
1977	1910[c]	3500	5410
1987	1920[c]	4300	6220
Change 1977–87	10	800	810
Change (%)	0.5	18.4	15.0
Toronto			
1983	364	740	1104
1988	408	925	1323
Change (%)	12.1	24.9	19.8

[a] Does not include construction.
[b] Includes agriculture; mining; electricity, gas, water; construction.
[c] Includes primary sector.
Source: Daniels, 1990

are international financial centres (IFCs) and Paris, Toronto and to a lesser degree Melbourne are regional financial centres; all are major metropolitan areas either within their own national systems or at the international level. The data for these and other cities not included here (London, Vancouver, Barcelona, Seville) show that absolute increases in employment in service industries have been taking place during the 1970s and 1980s at a time when the share of manufacturing employment has been declining in absolute and relative terms. There are of course exceptions, such as Toronto, but the rate of growth of manufacturing employment has been half that for service industries. This is not uncommon in medium and small metropolitan areas that are still expanding at the expense of other cities and regions within their national economies but are yet to be really competitive at the international level with the likes of New York or Tokyo. In Paris and the Île-de-France the decline of the manufacturing sector continues apace while the growth of service industries tends to be slower than in smaller cities. In the case of the smaller cities this may be symptomatic of a catching-up process as they begin to attract functions linked with the internationalization of service economies; but it may also reflect the continuing demand for population-related (consumer) services since these cities are still experiencing significant population growth while the very largest are losing population. The apparent deceleration in the growth of services in the largest cities is therefore relative, with producer services now leading the way in the restructuring process.

Comparable data for large cities in the less developed economies are difficult to assemble. Very limited information for Dakar reveals a large discrepancy between the population available for employment and those actually economically active (Daniels, 1990). There are some 200,000 to 250,000 formal jobs in a city of 1.8 million: a rate nearer to 50 per cent would be normal for cities in the developed world. According to surveys undertaken in 1976 (by Banque Mondiale) and in 1980 (by the Plan Directeur d'Urbanisme), between 70 and 85 per cent of the jobs in Dakar were in service industries. However, some 41 per cent were in the informal sector in 1980, a decrease from 45 per cent in 1975. This suggests that the formal, 'modern' sector of service employment has been increasing in Dakar, thus providing a more secure base for the city's economy city's economy which may, ultimately, help it to secure greater involvement in the international economy.

It is difficult enough to assemble comparable statistics on the broad structural shifts in employment in cities around the world. Even more problematical is any attempt to disaggregate these data to identify any differences in the behaviour of subgroups within the services. Not only are there variations in the classifications used by different countries, but city-level data are often not available, are incomplete in their coverage, or have only been collated during recent years. Some cities rely on the statistics from national population census returns while others gather their own information. The data in table 5.2 on the structure of service industries and their average annual growth must therefore be treated very cautiously; they provide broad indications only. In the late 1980s almost three out of every four metropolitan city jobs were in service industries (68–79 per cent). Distribution services are the principal source of service industry employment but they have tended to expand more slowly than services that are not population related, i.e. producer services. As a general rule, the larger the city, the greater the share of the producer services in total employment and service sector employment. Indeed, these services consistently emerge with the highest average annual growth rates over the decade spanning the mid 1970s to the mid 1980s; however, although absolute changes have been largest in the world cities such as New York, the average annual rates have been higher in the emerging international cities such as Toronto and Melbourne. The share of employment in public sector services seems to be connected with the status and function of cities: national administrative centres have a higher proportion of public services (London, Tokyo).

It is useful once again to contrast Dakar with the cities in the advanced economies. The size of the informal sector ensures that the majority of employment is provided by small enterprises engaged in the production of indigenous, traditional services. The range of activities involved is considerable, but most are characterized by low barriers to entry into the market, a dependence on resources available on or near metropolitan Dakar, and the lack of importance of educational qualifications. Control and regulation are a formidable problem if a more balanced structure of service industries is to be achieved.

Table 5.2 Changes in the structure of service industry employment, selected metropolitan areas

City	Distributive	Producer	Mainly consumer	Public	Non-profit	Total	No. of jobs (000s)
			Percentage of total employment				
New York 1987	27.7	21.9	4.2	15.1	10.4	79.3	5972
Av. annual Ch. (%) 1977–87	-0.30	3.77	2.10	1.42	5.10	1.97	
Île-de-France (Paris) 1985	4.3[a]	18.2	45.3			67.8	2093
Av. annual Ch. (%) 1975–85	0.70	0.20	1.89			1.31	
Melbourne 1986	22.9	21.6	4.6	10.3	16.6	76.1	195
Av. annual Ch. (%) 1976–86	1.05	5.44	3.53	1.73	4.06	1.49	
Toronto 1988	15.5	21.3	13.3	6.6	11.9	69.7	925
Av. annual Ch. (%) 1975–85	2.10	3.30	2.40	2.10	2.00	2.50	
Tokyo 1987	36.2	10.4		25.7[c]		69.1	4300
Av. annual Ch. (%) 1977–87	4.06[b]	2.22		0.32		2.28	

[a] Data classified into three groups: transport, commerce, services.
[b] Includes eating and drinking places.
[c] Other services.

Source: Daniels, 1990

Table 5.3   Growth of selected producer services, New York metropolitan area, 1977–1987

Producer service activity	Absolute growth 1977–87 (000s)	% change 1977–87	% of all service industry employment 1977	1987
Finance, insurance, real estate	237	40.8	12.3	13.7
Business services	230	70.0	7.0	9.3
Legal services	43	77.4	1.2	1.7
Membership organizations	6	10.3	1.2	1.0
Miscellaneous professional services	33	38.0	1.8	2.0
Total	549	49.5	23.5	27.7

Source: Scanlon, 1990

Although producer services have been the principal source for absolute job growth during the 1980s in some of the world's leading cities, different types of producer service have not been expanding at the same rate. In New York, for example, the bulk of the growth was generated by the activities included in finance, insurance, real estate (FIRE) and business services (table 5.3). These contributed more than 85 per cent of the job increases and together comprised 23 per cent of service sector employment in New York in 1987 (19.3 per cent in 1977). Business services employed 558,000 in 1987 with an annual average growth of 5.5 per cent per annum during the preceding decade. Even higher rates of annual growth were recorded by securities and commodities traders at 8.5 per cent (absolute increase 1977–87 was 75,000). Within the Square Mile of the City of London total employment increased by 6.4 per cent between 1981 and before the Stock Market crash in October 1987. This compares with 11.3 per cent for services and 34.3 per cent for producer services (including finance) – an annual rate of between 5 per cent and 6 per cent.

Another index of the growing importance of producer services in metropolitan economies is their contribution to total metropolitan incomes and the contribution of exports to metropolitan income. It is difficult to measure the contribution of service industries to metropolitan region growth because of the acute data problems, but Drennan (1985) has produced estimates for the New York region. Total regional income from the export sector (manufacturing, transport and communication, wholesale trade, FIRE and services) in 1958 was $26b, of which $10b was derived from exports, with FIRE and services contributing 45 per cent of this total. By 1983 regional income had risen to $48b, the export component to $19b and the FIRE and other services share to 52.8 per cent. Securities dealing alone contributed 17 per cent of the export component of regional income in 1983 (only 7.1 per cent in 1958). It would be useful if, in relation to the theme of this book, the international component of export income could be identified (including changes over time). Inevitably this remains difficult, but Warf (1987; 1991), for example, notes that advertising firms located in Manhattan derive roughly 51 per cent of their income from foreign clients. In the larger New York Standard Consolidated Area, exports comprised 26.8 per cent of the total receipts of consulting and public relations services and 17.8 per cent of receipts for engineering and architecture services in 1982 (Warf, 1991; see also Noyelle and Peace, 1991; Drennan, 1992).

Impacts on the Urban Property Market: Offices

Sectoral restructuring towards services and the associated employment growth, especially in international and national market financial, professional and business services, has had a major impact on property markets in cities. The effects have been experienced in residential, retail and industrial property development (see for example Thrift and Leyshon, 1992; Thrift, Leyshon and Daniels, 1987) but have perhaps been most clearly displayed in the market for office buildings. These are not simply required to accommodate additional jobs but must be available in configurations able to meet increasingly exacting requirements, enabling the use of information technology for the production and distribution of services. The design and internal configuration of individual office buildings must incorporate a range of special facilities which allow occupiers to link easily with the full range of telecommunications services, from local area networks (LANS) to satellite links. The flexibility to

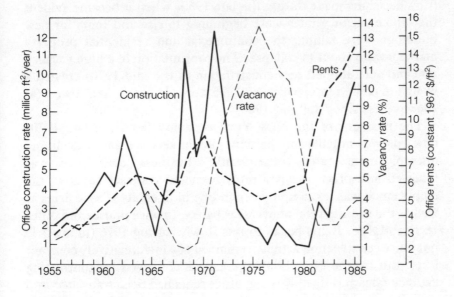

Figure 5.1   Manhattan office construction, rents and vacancy rates, 1955–1986
Source: Warf, 1987

reconfigure the IT systems within office buildings as updated equipment is acquired or as organizational utilization of IT is modified is also vital (Black, Roark and Schwartz, 1986).

The construction of new office space in cities tends to occur in cycles as the development industry tries to match the supply of vacant office space with the anticipated demand from occupiers (Barras, 1983; Bateman, 1985). Business and office development cycles do not always coincide, however, because at least three to four years elapse between a decision to construct a new office building and its availability for occupation; much can change during the interim period (see for example Barras, 1983).

The property market response in Manhattan provides an excellent example of the way in which the rapid expansion of information services in New York has affected the demand for good quality office space (Warf, 1987; 1991). We have seen how in the mid 1970s the New York economy was on the brink of collapse. These circumstances are reflected in the precipitous decline from 1970 in the construction of office space in Manhattan (figure 5.1). This arose from decisions made during the late 1960s when it became evident that office vacancy rates were beginning to rise and rents for new buildings were falling; the commercial and residential property markets were about to collapse. The continuation of a high vacancy rate and very limited new construction in the mid 1970s created a trough in office rents, which by 1975 were almost at their lowest (at 1967 prices) since the mid 1960s.

The recovery of the New York economy from the late 1970s onwards was notable for the new jobs in services and a declining unemployment rate; consequently, business confidence was renewed and office vacancy rates began to contract again as the empty stock was taken up by expanding indigenous service firms as well as the in-migration of overseas banks, finance houses and other service MNEs. Rents began to rise slowly during 1976 (figure 5.1) but office construction activity remained at a low, relatively constant level. But by the early 1980s the signals conveyed by contracting vacancy rates and steadily rising office rents had been translated into renewed construction activity, leading to a steep increase in new office development and a very sharp increase in rents as demand outpaced the supply of vacant office space, particularly between 1980 and 1982. The boom in the office market continued into the mid 1980s as bullish international financial and professional service markets encouraged more service MNEs to expand into new regions around the world. But the bubble burst in October 1987 when the New York Stock Market triggered a sharp fall in share values in other major stock markets. The world economy has since moved into a downward phase in the business cycle, culminating in a recession which has been accompanied by very large decreases in, for example, the volume of international financial transactions, which in turn has resulted in labour shedding and even the withdrawal of some MNEs from direct representation in overseas markets. Thus, a new downward phase in office construction has occurred in Manhattan: office vacancy rates have started to climb again to levels paralleling those of the mid 1970s. Rents have also fallen and the appearance of another trough in the office development cycle has been initiated.

To some extent, the timing and duration of office development cycles will vary between cities in different parts of the world, but the way the market works is much the same everywhere. The West End

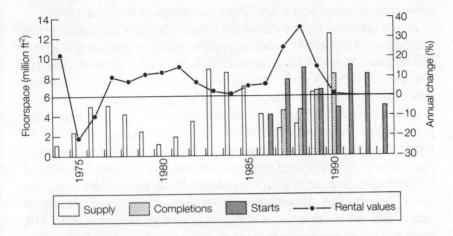

Figure 5.2   Changes in the supply, completions, starts and rent values for central London offices, 1974–1993
Source: compiled from Debenham Tewson Research, 1990, 1991

and City office markets in London have behaved in much the same way as New York and for very similar reasons throughout the 1980s (Duffy and Henney, 1989; Fainstein, 1990; Daniels and Bobe, 1990). Fluctuations in the supply of office space immediately available for rent in central London between 1974 and 1990 have followed a seven or eight year cycle: the peaks were in 1977, 1983 and 1990 (figure 5.2). The changes in annual rental values broadly reflect the peaks and troughs in supply: increases tend to be greatest when the supply of office space is at its lowest, for example 1987 and 1988. As in Manhattan, the completion of new office buildings in central London in the late 1980s did not coincide with the boom in employment growth in 1986 and 1987 following 'Big Bang'. Completions peaked in 1988, some three to four years after the plans were made to construct the buildings, when the overall supply of space was lowest and annual rent changes reached a staggering 30 per cent. The distribution of office construction starts between 1986 and 1990 clearly demonstrates the lead and lag between starts and completions. As more new office space was completed and available for lease in 1990 and 1991 in central London, the slowdown of the frenetic employment growth of the second half of the 1980s was well established and rents were falling back. We can therefore expect fewer construction starts for office buildings in the mid 1990s.

As the internationalization of the financial system has transformed capital into a commodity that is readily tradable around the globe, the ebb and flow of office construction in cities has been affected. The principal nodes in the circulation of capital are the international financial centres (see chapter 4). The equity markets are the traditional focus for business but investment in office development projects has also greatly increased. This can be understood best in relation to the stimulus to demand for locating producer services in the international financial centres provided by the combination of deregulation and innovations in information technology from the late 1970s onwards. Investment is therefore not confined to national banks and other financial institutions; foreign investment in UK property, for example, increased from £200m in 1985 to an estimated £3.1b in 1989 (Hugill, 1990). Almost all this investment was channelled into office construction in central London (the City and the West End). During 1987–90 approximately one-third of all investment in central London came from overseas (Japan, United States, Scandinavia, Kuwait and Australia, for example), reaching as high as 40 per cent of the total investment during 1988–9 (Hugill, 1990).

The diversity of the sources of international investment in central London office development is shown in figure 5.3. The substantial involvement of Japanese investors is notable and has continued to grow during the early 1990s. They are primarily interested in the most prestigious (trophy) office investments; the potential for a continuation of the flow of foreign investment into London, New York or Tokyo is therefore determined to some extent by the quality of the office buildings available and/or the extent to which the existing stock of buildings is being replaced or refurbished. Apart from possible problems of attracting international investment, failure to modify the stock of office buildings would render most of the world cities uncompetitive with respect to the requirements of international business and professional services.

It is no longer adequate to construct office buildings that are essentially a 'shell' which individual occupiers must fit into or have minimum scope to organize in the way that fits with their needs. They now expect a number of features to be standard in any new or

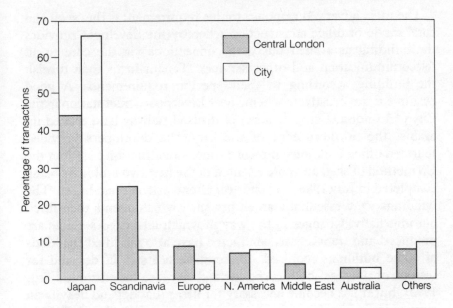

Figure 5.3   Foreign investment in office buildings/development in the City and central London
Source: Richard Ellis Research, cited in Blanden, 1990

refurbished office building, such as ancillary services (flexible and high-capacity ducting for telecommunications and IT equipment, backup power supplies), high-quality lighting and air conditioning provided in a way that allows maximum flexibility in the configuration of working spaces, and greater emphasis on the quality of the working environment in the building. According to Duffy (1986) 70 per cent of the cost of office construction in 1965 was derived from the building shell; in 1985 this was down to 40 per cent, while the costs arising from the provision of building services had doubled from 20 to 40 per cent. The increased emphasis given to the quality of the working environment in office buildings means that the costs for landscaping have also doubled from 10 per cent in 1965 to 20 per cent in 1985. Thus 'it is no longer sufficient to provide a basic building whether new or refurbished, and expect it to let – it is vital to offer a well-conceived product where a high degree of consideration has been given to the design for modern occupational needs' (Peach, 1985, 34; see also Leyshon, Thrift and Daniels, 1987a; Williams, 1992).

The most advanced response to this requirement is the 'shell and core' mode of office construction whereby the developer provides the building as a shell but with connections via the core to all telecommunication and other services. Tenant firms then furnish the building according to their specific requirements. A good example is the Broadgate scheme near Liverpool Street station in the City of London. Using 29 acres of disused railway land tucked up against the northern edge of the City, the developers provided fourteen office buildings in record time. Construction based on the US method of shell and core resulted in the first two buildings being completed in July 1986, just one year after work on them began. The buildings were essentially an empty shell which clients then fitted out internally. Changes in the way in which advanced services are produced and transactions conducted have also modified the kinds of office buildings required. Skyscrapers are still in demand for reasons of prestige, both corporate and city (see for example Ford, 1992), but it has become necessary for firms that depend heavily on electronic dealing, for example, to weigh up carefully the advantages of prestige against the working environments required by dealers, i.e. large dealing rooms housing all the electronic support equipment required but also permitting visual communication between dealers in an environment frequently requiring quick decisions and as much up-to-date information as possible. For some kinds of advanced services, especially in finance and securities, 'groundscrapers' (low rise, large individual floor areas, often as many floors below ground as above) are now functionally more appropriate than skyscrapers (high rise, small individual floor areas) (Williams, 1992).

The combination of pressure of demand from international service firms and the requirement for office buildings with very different specifications has transformed the built form of the CBD. The best example is central London and, most notably, the City of London. Here the highly irregular pattern of streets, plot sizes and controls on the height of office buildings had, until the late 1980s, created a skyline punctuated by a very small number of modest skyscrapers (e.g. the National Westminster Bank Tower or the Lloyd's Building) and a patchwork of individual office buildings of limited scale and floorspace. The requirement of the financial institutions, in particular, for buildings with extensive individual floor areas (3000–5000 m² or 30,000–50,000 ft²) for dealing rooms, for example, required buildings that extended across existing sites and sometimes replaced office buildings that were no more than 20 years old; the London Wall area is a good example. Consequently old building lines and street patterns have been modified in the interests of retaining and attracting international service functions. The 'air space' above the main railway termini (such as Liverpool Street, Charing Cross or Holborn Viaduct) in and around the centre of London has also been used for large but extensive rather than tall office buildings and retail developments which have also added further to modifying the city centre landscape. As Williams (1992, 250) has observed, not all the changes are necessarily positive in that the 'challenge of building big and bulky has yet to be matched by substantial and inventive external design, or detailing that relates to richness and craft. Sadly most of the façades which are now being revealed are drearily banal.'

Of course, not all the service firms requiring a location in the major global cities require state-of-the-art intelligent office buildings. The smaller international firms may not have the resources to pay the rents demanded, while many firms are largely concerned with serving national markets or have links with clients that are international market services. This has resulted in a two-tier office market in many major cities. This may take the form of a distinction between firms able to pay premium rents and those able to afford lower rents for generally less well specified, often older, office buildings. But rents tend to be determined by location as well as building quality and prestige. Some service firms will prefer a core (or first-tier) location at any cost: the investment banks and securities houses, insurance firms, corporate legal services and commodities brokers. But others can be more footloose about their location choices and are able to occupy office space in marginal locations. These are often locations that ten or twenty years ago would not have been considered acceptable; midtown Manhattan in New York, or areas on the edge of the City of London or the West End. Accountants, solicitors as well as investment banks have been forsaking the traditional locations in the City of London or on Wall Street for locations such as Broadgate or Battery Park City that attract lower rents but are still easily accessible and may indeed offer office buildings that are at least as good as if not better than (as at

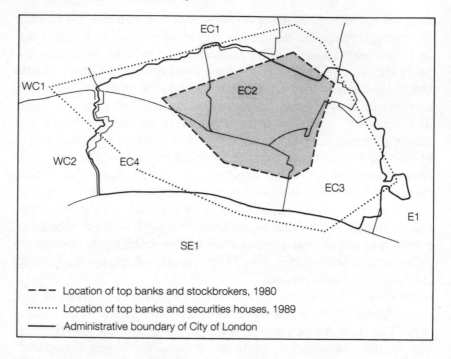

Figure 5.4 The expansion of the functional City, 1980–1989
Source: information in Jones Lang Wootton, 1990

Broadgate) those in the long-established core locations in the cities. One of the consequences of this process is an extension of the functional area of the City of London (figure 5.4) or the financial district of Manhattan. The functional City in 1980 was almost entirely contained well within the Square Mile (delimited by the administrative boundary of the City of London: see figure 5.4); by 1989 its limits had been extended, especially to the west and to the north (see Moss, 1991 for data on the location of foreign bank offices in downtown and midtown Manhattan).

Changes in the Location of Services Within Cities

Although there is evidence for a growing concentration of employment growth in finance and legal services in cities like New

Table 5.4 Concentration of growth of selected producer services in New York City, 1979–1987

Producer service activity	NY Met. Region change 1977–87	NY City change 1979–87	NY City share (%)
Finance, insurance, real estate	237	105	44.3
Business services	230	69	30.0
Legal services	43	26	60.4
Miscellaneous professional services	33	7	21.2

Source: derived from Scanlon, 1990; Noyelle, 1989

York throughout much of the 1980s (table 5.4), there is also evidence for some locational redistribution. In New York the CBD (Manhattan) was the scene for an increased share of FIRE and business services in all employment between 1977 and 1986, but all other services constituted a smaller share of the total when compared with the equivalent distribution in 1977 (Warf, 1987). The percentage growth has been lower in the CBD for all services by comparison with the NY region as a whole, with the sole exception of FIRE employment which increased by 41.9 per cent. Such was the strength of this performance during the period of New York's 'renaissance' that the overall change in CBD employment (16.8 per cent) remained ahead of the region (14.6 per cent). The deconcentration implied by such statistics is therefore rather selective (as we have already seen in chapter 4) and it is not unique to New York.

Using a rather different measure based on tradable and local service employment, there are indications of a limited redistribution within the Toronto metropolitan area (table 5.5). The share of tradable services in the core had fallen below 50 per cent by 1988, with the fringe and the suburbs increasing their share. While the suburbs retained a constant share (73 per cent) of local service employment there was a slight redistribution from the core to the fringe areas around it. Some 57 per cent of the business services found in the metropolitan region of Barcelona are located in the City of Barcelona (in the centre of the metropolitan area) (Baro et al., 1989). The proportion is even higher (88 per cent) for well established, stable business service firms. In the remainder of the

Table 5.5   Changing location of tradable and local service employment, Toronto, 1983 and 1988

| Location | Employment (000s (%)) | | | | Growth (%) | |
| | Tradable[a] | | Local[b] | | Tradable | Local |
	1983	1988	1983	1988		
Core	124 (50)	150 (47)	67 (14)	81 (13)	20	21
Core/fringe	23 (9)	36 (11)	64 (13)	83 (14)	57	30
Rest of met.	100 (40)	136 (42)	367 (73)	439 (73)	36	21

[a] Tradable services can be exported outside the metropolitan area and therefore attract income to it. Included are: education (post-secondary), hotels and lodging, film, producer services, legal and finance, insurance and real estate services (which are in the zone of the metropolitan area), federal and provincial governments.
[b] Local services are not traded outside the city and therefore do not bring in new wealth. Included are: distributive services, non-profit services, education (post-secondary 65 per cent), all other consumer services, bank and trust company local branches, legal and other financial services (not in the core), local and regional (metropolitan) government.
Source: Richmond, 1989

metropolitan area the most important concentrations of business services are in municipalities which have a well developed manufacturing base or an administrative function (such as a district capital). Centrality, perception and social evaluation, the rise of economic activities and infrastructure services are all important and lead to business services looking favourably upon the core of the Barcelona metropolitan region for the location of their activities.

The relocation of offices, mainly in the service sector, from central London has been a regular event since the mid 1960s (figure 5.5). Most of the moves have been to suburban centres such as Croydon, Hammersmith, Wembley or Ealing, with the numbers rising throughout the 1970s when they averaged 7000 jobs per year. During this period government policy was largely responsible for encouraging relocation, but the annual exodus fell back sharply during the 1980s following the removal of planning controls and the demise of agencies such as the Location of Offices Bureau. By 1990, however, the mobility of firms had returned to the pre-1980 levels. Rents for good quality office accommodation more than doubled in central London between 1985 and 1989, reflecting the competition for office space amongst international business, financial and professional services. The Big Bang in 1986 exacerbated the

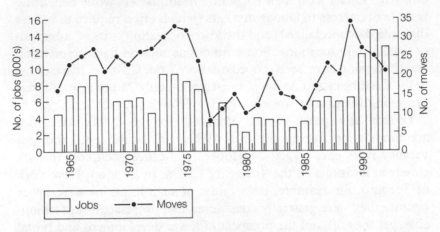

Figure 5.5 Office relocation from central London, 1964–1992
Source: compiled from data in Jones Lang Wootton, 1991

problems because it coincided with a shortage of immediately available and suitable office accommodation. With a steep rent gradient outside London, companies saw an opportunity to save on operating costs and improve accessibility to labour, especially skilled and clerical staff that had been in short supply in central London during the boom years. Information technology may further support the case for future relocations although, relative to the overall size of the central London office market and labour force, the impact of relocation activity should not be overestimated.

A feature common to all the cities that have experienced some relocation of service functions is the selectivity of the process. In general, front office staff involved directly in corporate decision making and control (a group increasing in size as service firms become larger as a result of mergers and similar activity) remain concentrated in the CBD locations where their information needs or access to specialist markets and workers are most easily fulfilled. But staff involved in data processing functions or marketing have less need to be located centrally and are increasingly relocated to suburban centres within the same city or to smaller, regional cities (Stanback, 1991; Moss, 1988). These so-called back offices are able to take full advantage of information technology to maintain full and effective contact with their corporate headquarters while benefiting from easier access to labour markets (which often require to be less diversified or specialized) and the lower operating costs of suburban or smaller city locations. Some functions such as data preparation and processing have been moved offshore from US cities to locations in the Caribbean and Asia where wages are only 25 per cent of those prevailing for similar work in the US (Howland, 1992).

During the peak of the demand for good quality office accommodation during the mid 1980s, escalating rents and low vacancy rates encouraged developers and office occupiers to look closely at locations on the 'fringe' of CBDs. In London, New York or Toronto, for example, land values at such locations were lower because they were generally considered inappropriate or unfashionable, yet they offered the prospect of lower development and rental charges without seriously sacrificing access to the CBD corporate complex. As an alternative to locations further afield in the metropolitan suburbs and beyond there seemed much to commend such developments. An early example of the attraction of this approach is La Défense on the western side of central Paris which was planned in the 1960s as a large-scale office centre supported by substantial up-front public investment in carefully planned transport infrastructure (Bateman, 1985). Its role was to divert demand for good quality office space in the centre of Paris so that the conflict between modern office buildings and the historic urban landscape would be minimized without necessarily undermining the key national and international role of Paris as a location for the offices of service and manufacturing MNEs. After more than twenty years of careful development La Défense is an undoubted success.

More recent examples may be modelled on the La Défense idea but have been implemented in a very different way. Developments such as one Canadian Place (Toronto), Battery Park (on the south-western fringes of downtown Manhattan on land reclaimed from the Hudson River) and Canary Wharf in London's Docklands represent an alternative approach to providing high-quality office space at competitive rents in marginal and unfashionable locations (Zukin, 1992). Ultimately, all these projects have been instigated or taken over by Olympia and York Development Limited, a Canadian property company.

Canary Wharf is about 4 km (2.5 miles) to the east of the City of

London and is the largest single development of new office space in Europe: 1 million m^2 (10 million ft^2) net of offices, 50,000 m^2 (500,000 ft^2) of retail and leisure facilities and a 400 bedroom hotel with conference facilities (Daniels and Bobe, 1991). The plan for Canary Wharf was hatched in the mid 1980s as the demand for central London offices was moving towards its peak. Development opportunities in the City and in the West End (central London) were limited and the sites that were available were extremely costly to acquire. As we have seen, developers were not only faced with high land or property acquisition costs but were also expected to provide high-specification, intelligent office buildings suitable for use by international financial, professional and business services. Such structures would only be viable at very high rents; some clients would accept them for the sake of the prestige and agglomeration economies of a City or West End address but many others would look for lower-cost accommodation in other international business centres elsewhere in Europe.

The key to Olympia and York's development philosophy is that by purchasing very low-cost sites in unfashionable locations, very high-quality office accommodation can be provided which will be attractive to 'blue chip' service and manufacturing MNEs. London's Docklands, long abandoned by port functions that have moved downstream to Tilbury, were just such an opportunity. 'They had the added attraction that, because they were part of a government designated enterprise zone, new developments were exempt from business rates until April 1992, there were 100 per cent capital allowances available on buildings where construction started prior to 1992, and planning controls were limited, more flexible and speedier. Olympia and York were clearly making a great leap of faith that the very high-quality office that they could provide at lower rents than those prevailing in the City would compensate for the disadvantages of the location. This contradicts long-standing developer attitudes which essentially dictate that all that matters for a successful office development is location, location and location. Some of the disadvantages confronting the development in Docklands include the low esteem with which the area has always been held as an office location within London, the consequent difficulties of attracting suitable labour, and most important of all the inadequacy of the transport infrastructure for coping with the

60,000 office workers (plus visitors and other traffic) converging on the area if and when all the office space is occupied.

Transport considerations have therefore been at the forefront of the debate about the likely success or failure of Canary Wharf (Church, 1990). The Docklands Light Railway, (DLR), which links the City (Bank Station) to Docklands, was planned before the Canary Wharf project was conceived, and although its capacity has been expanded to 24,000 passengers per hour it will still be inadequate to meet peak hour demand (apart from the continuing unreliability problems that have dogged it from the beginning). An extension of the DLR to Stratford (thus providing a link with British Rail services) will help to improve access, along with the planned construction of a new underground line into south-east London, the Jubilee Line, which will include a station at Canary Wharf. But not all office workers, visitors and others users will want to use rail services to reach Canary Wharf; the road system remains totally inadequate with respect to access from the City and other parts of London and to the internal circulation and parking facilities within the Docklands area.

Perhaps more seriously for Olympia and York, the office space at Canary Wharf started to come on stream in 1991 at a time when, for the reasons discussed earlier in this chapter, there was surplus office space in London, and the over-supply seems likely to increase well into the mid 1990s. Not only has international demand for office space in London slowed down, but established London firms have been shedding staff or not growing while some overseas firms have been withdrawing from London or cutting back on staffing levels, thus reducing the pressure to search for new or additional accommodation. By late 1991 agreements had been reached for the take-up of some 300,000 m² (3 million ft²) of office space at Canary Wharf but the cost to Olympia and York has been considerable since, in some cases, they had purchased the leases of buildings previously occupied by their new tenants elsewhere in London. Most of the tenants are North American companies that are already familiar with Olympia and York's business philosophy in New York or in Toronto. Some major British service MNEs, for example in advertising, were almost lured to Canary Wharf during 1991 but withdrew at the last hour. Clearly, the scheme requires a number of anchor or blue chip tenants to raise the confidence of the wider business community about the acceptability of the location as an alternative to the City, especially for back office functions. Following problems in North America that it endeavoured to separate from its scheme in London, Olympia and York finally succumbed to the administrators in May 1992 and relinquished ownership and control of the Canary Wharf project, owing large sums of development capital to banks in the UK, Canada and the United States.

While the future of the office space on Canary Wharf remains uncertain, as does the overall completion of the project, the outcome demonstrates some important points about the role of services in the contemporary world economy. Firstly, it has been suggested that the impetus for Canary Wharf came from the demand for high-quality office space generated largely by international financial, professional and business services rather than national firms. Secondly, such was the scale of the project that finance capital from domestic (UK) sources was never going to be adequate; it depended upon international flows of capital investment of the kind made possible, for example, by deregulation, innovations in financial instruments and information technology. Thirdly, the circuit of supply and demand for office space, infrastructure and human resources in cities is therefore increasingly mediated by a select group of (financial) services that both enable the process of development and change and participate in the benefits. Fourthly, the close intertwining between developments like Canary Wharf and the world of internationalized services, but also the more general process of growth and change in nations and cities, increases the vulnerability and the uncertainty about success or failure. It has become more difficult to isolate bankers' assessments of the ability of Olympia and York to meet its debt provisions on a large development in London from the assessments made by bankers on its property portfolio in New York.

Capital is footloose and producer services are footloose: information technology has seen to that. The expansion of services into the international marketplace has therefore introduced a greater degree of flexibility, and ultimately competition, into the global urban system than was the case in the past. As the experience with Canary Wharf has shown, it has also made the outcome of large-scale planning and redevelopment within cities a hostage to external international factors over which they can have limited control.

It remains to consider whether some of the issues raised by this experience, together with some of the trends outlined in the earlier chapters, give cause for optimism or pessimism for the future role of services in the world economy. The questions raised in the final chapter could well have been used as the point of departure for the selection of subjects covered in this book; in the event they may provide a focus for discussion about the things which we should not take as given on the basis of recent trends and some of the questions that might be addressed in future analyses of the significance of services in contemporary space economies.

SIX

Services in the World Economy: Some Reflections

Service industries have enabled, and themselves become participants in, world trade. In so far as they have helped to underpin the production process or enabled circulation of capital and labour, their historical contribution to the growth of the global economy should not be underestimated. More recently they have actively participated in the growth process by creating new or diversified products that have resulted in increased consumption in anything from tourism and leisure to completely new ways of raising finance capital to support corporate growth or production strategies. But by the very nature of the global economic environment there are spatial variations between nations, regions and cities in factor endowments that ensure that the demand for services, whether in the form of knowledge, skills or information transfer or in the form of access to different (or better) educational services or more exotic tourist destinations, will continue to expand during the remainder of the 1990s. It seems unlikely, however, that the various subsectors within the services will, for example, grow at the same annual rate as in the 1980s or that the balance of employment or contribution to GDP will remain the same. Advances in productivity, further innovations in information technology or the potential for modifications to the geography of comparative advantage as the less developed countries become proficient in the production of business services will see to that. Of course the factors that will contribute to the future form and structure of services in the global economy and their consequences for the global urban system, in particular, are numerous and worthy of a completely separate analysis. Space precludes this here; suffice to point to some of the unresolved issues that will have a bearing on the part that service industries will play in metropolitan and global development during the relatively few years left in this century.

Services in the 1990s: Victims of the Decade of Optimism?

The globalization of services during the 1980s reflected great optimism about the universality of demand and the inevitability of the imperatives created by information technology. But just as local or regional economies are vulnerable to fluctuations in the business or the economic cycle, so too is the global economy. It has been widely expected that the events of the 1980s will provide a launchpad for a decade in the 1990s of accelerating global investment in services: capital will move easily across national borders, using sophisticated telecommunications, in the search for the best returns in different national markets that are never fully synchronized in relation to macro-economic cycles. For once, however, the economic downturns around the world may vary in depth and duration but seem rather more closely synchronized in the early 1990s than has been the case in the past. This has retarded the energetic international investment that was the hallmark of much of the 1980s, fuelled by the disillusionment of some institutional investors with the results of their forays into overseas markets. It is worth speculating whether the synchronization of phases in the economic cycle around the globe is not in some way linked with the advent of high-speed telecommunications which make it difficult for places to hide from the ripple effects of a stock market crash or of an economic recession.

A good index of the less optimistic outlook for services in the 1990s is the condition of the property market in the global cities, whose economic fortunes during the 1980s have been closely connected with the internationalization of services. For example, between 1984 and 1989 banks in the US channelled 60 per cent of the increase in bank loans into real estate (much of it to finance new office buildings in the diversified service centres such as New York and Chicago). US savings and loans institutions also invested heavily in commercial property.

Consequently, the volume of office space constructed in the late 1980s far exceeded foreseeable demand even before the onset of the US and the more general global recession; empty office buildings and tumbling rents have become the norm, and will remain so for some time. Defaults on the bank loans are rising as office development companies go bankrupt, so reducing the ability of the banks to issue new credit, and this in turn inhibits growth in the economy in general. Much the same is true in the UK, where the international involvement in London's office market escalated in line with the deregulation of the financial markets in the City of London but generated high office rents and a large volume of new office space to replace a large stock of obsolescent space. The recession, the caution instilled by the stock market crash of late 1987 and the cutbacks in staffing or even withdrawal altogether from London have led to rising office vacancy rates and falling rents. Even the cities on the European mainland, such as Frankfurt and Paris, that are expected to benefit from the strong economic growth following the creation of the internal market in 1992 have not been exempt.

The growth of employment in business, professional and financial services which consolidated the powerful position of Tokyo, New York, London and a handful of other cities in the operation and control of the world economy throughout the 1980s cannot be taken for granted in the 1990s. Historically, services have been less vulnerable to fluctuations in economic growth than manufacturing or extractive industries. Recent evidence indicates that this is no longer the case, and that the competition, specialization and diversification that created a bullish attitude towards job creation in many of the services during most of the 1980s was over-optimistic and would inevitably be followed by contraction to more realistic levels of employment. This has indeed been happening, although it may also have been helped along by the global economic recession. A case in point is New York, where private sector job losses, most notably in services, have been continuing in the wake of the 1987 stock market crash. In 1990 alone the private sector shed 51,000 jobs in New York city, the biggest annual loss since 1975. The commercial banking sector is particularly important to the NYC economy: it employed 108,300 at the end of 1990, but 10,000 were lost in the first three months of 1991 (almost as many as in the whole of 1990) (Zagor, 1991). A merger between Chemical Bank and Manufacturers Hanover led to a further 5900 job losses in commercial banking later in 1991.

These and other job losses in services threaten to precipitate yet another fiscal crisis for New York, since the city must cope with losses of tax revenue from highly paid workers, declining property values (and taxes) and declining support for restaurant, entertainment and related services. For every banking job lost an additional 1.4 jobs will be shed by other business, and lost taxes will amount to some £3976 per job cut when secondary impacts on property and other areas are included. But while the earlier fiscal crisis in New York was generated by restructuring of its economy from manufacturing to financial and business services, the present crisis reflects a search for equilibrium in an overheated service sector. Thus, New York remains a major financial centre which continues to attract foreign banks, for example. The difference is that their investment in New York office space and plans for employment growth will be much more modest. It can therefore be argued that although the potential for decline is less drastic than in the 1970s, the consequences may be just as severe because the loss of highly paid service jobs will have a disproportionate impact on New York's economic welfare during the 1990s.

Another symptom of the changing climate can be found in the market for internationally traded services. There is now a growing willingness amongst firms that have traditionally been competitors to establish alliances that will help to retain their share of world markets. The Chicago Board of Trade and the Chicago Mercantile Exchange, for example, were pre-eminent in the trading of futures in the second half of the 1980s but their share of the market was shrinking steadily as other futures exchanges opened around the globe. With less than 50 per cent of the market between them in 1990 and much of the new business in futures going to other centres such as the London International Financial Futures Exchange (Liffe) or the Marche terme international de France (Matif) in Paris, the two Chicago exchanges have put aside their long-standing intense rivalry in order to protect their common interest (*Financial Times*, 1991c). Apart from launching a series of new products with international appeal, jointly developing an electronic trading card to improve exchange and regulatory surveillance and consolidating their foreign representative offices in London and Tokyo, the two exchanges have joined Globex, an after-hours electronic trading system. This allows them to capture overseas business which Chicago's hours do not permit, and some of their non-US competitors have yet to sign on for the system although they are being set up as partner exchanges, (e.g. Liffe, Matif, the New York Mercantile Exchange (Nymex) and the Singapore International Monetary Exchange (Simex)).

Service-Dominated Economies: How Desirable?

The continuing trend for restructuring of national and metropolitan economies throughout the last decade (and indeed earlier) has also fuelled a long-running debate about the merits of service-dominated economic systems. A member of the board of British Invisibles commented in late 1991 that it 'really won't do these days to make judgements about the state of the British economy on the basis of "manufacturing output". One plainly cannot arrive at a sensible conclusion if one ignores services, which have acquired a greater significance than ever before' (*The Times*, 1991b). With over half of Britain's foreign exchange earnings accounted for by private invisible exports (such as insurance, banking, shipping, aviation, tourism) the argument seems uncontestable.

Of central concern is whether the decline of manufacturing employment or the shift in balance of trade towards imports rather than exports of manufactured goods (figure 6.1) signifies terminal decline of advanced market economies such as those of the UK or the US and their global competitiveness. Is not the shift to services just normal economic evolution (in line with the model outlined in chapter 1)? There is a certain 'romantic attachment to the supposed economic virility of manufacturing as against supposedly effete services' (*The Times*, 1991b), and there is a view that service industry should not be seen as a substitute for manufacturing because it is dependent on the latter and only 20 per cent of services are (in the case of the UK) tradable overseas. Both the UK (see figure 6.1) and the US are concerned about how the balance of trade can be brought

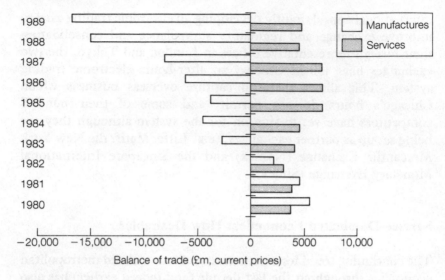

Figure 6.1   Balance of trade in services and manufactures, UK, 1980–1989
Source: Central Statistical Office, 1991

back into equilibrium. Even if there is rapid growth of exports by producer services, they actually represent a small proportion of overall services exports (the largest part is investment income – interest income and corporate retained earnings) and, indeed, there may be an increasing tendency to import services that are labour intensive and can be competitively provided by countries with lower labour costs.

Ideally, economies need to be strong in both services and manufacturing because of the value added in all sectors, but especially in manufacturing where more of the value added is retained in the country of production for employees, businesses and the national exchequer. But is such a distinction really relevant? Following the rise of the TNCs and MNEs, we should recognize that this distinction is a statistician's convenience, and nations should concentrate on generating the economic activity best suited to any competitive or comparative advantages that they possess without forcing an increasingly artificial distinction between services and manufacturing as the basis for formulating policies. As always in debates of this kind much depends upon the interpretation of statistics and the perceptions of the protagonists.

In an overview of both sides of the debate Perna (1987) distinguishes between the benign and the concerned view of the changing industrial structure of the US. Those taking the benign view (see for example McUsic, 1987) point to the fact that manufacturing's share of real GNP in the US has not declined; its share of jobs has been falling but this has been characteristic of the US economy for several decades; the shift to services has not retarded the average rate of productivity gain, nor has it reduced average wage levels; and just as there are plenty of opportunities within the US for the growth of services there must be at least as many opportunities for exporting them. The concerned view suggests that the facts are not so simple. Thus, defining manufacturing's real share of the US poses major measurement problems such as the sensitivity of the real GNP calculation to the particular base year that is chosen: the more recent the base year the smaller the manufacturing share of GNP (Perna, 1987). The decline of US manufacturing jobs is not just a continuation of a long-term trend; the decrease has been accelerating, with an annual rate between 1978 and 1985 nearly three times higher than the period 1968–78. While services productivity measurement remains a problem it is almost certain that faster growth of labour productivity in manufacturing has contributed to job losses while labour productivity in services has actually turned negative. This is significant because it drags down the overall average productivity at the same time as international competitors such as Japan, but also some of the developing countries, are improving their productivity levels very quickly (with the added advantage of lower wage rates than in the US).

Is the Globalization of Services Inevitable?

It is also necessary to consider the extent to which the trend towards globalization of services is inevitable. Here the messages are far from clear. It is very easy to get carried away with the concept without noting that only the very largest service enterprises in all sectors are either involved or actively contemplating global strategies; for the vast majority of services it is a process that they may benefit from indirectly by tapping into the local and national demand created by the behaviour of their larger competitors. That is the limit of their ambition. Even for the enterprises that have internationalized, their domestic markets remain a major anchor for their business success and profitability. It is doubtful that many service enterprises are truly global, i.e. they serve a wide cross-section of domestic and commercial clients using a fully integrated network of offices rather than a limited group of other multinational organizations.

Certainly, few banks believe that a global consumer banking network is feasible. One of the only exceptions is the New-York-based Citibank. While a number of British and French retail banks have withdrawn from expensive and loss making forays into overseas markets, Citibank has remained faithful to the concept of global banking. Other US banks such as Chase Manhattan are withdrawing from overseas locations, and Japanese retail banking is largely confined to California (in contrast to Japanese investment banking which is very much in evidence in the leading global financial centres). In mid 1991 Citibank served 34.5m households in 40 countries (including 24m in the US) through 1680 branches (551 in the US). Citibank also has 41m credit card holders (10m outside the US) (Lascelles, 1991). A measure of the commitment necessary to create such a large global presence and to aspire to a truly networked service is that the bank manufactures its own cash machines and has some of the most advanced credit card processing centres. In 1991 only the Japanese and US branches had cash machines linked to each other; a single global network is some way off. In order to ensure good quality control and to benefit from economies of scale, Citibank provides a limited range of banking services using standardized technology and banking premises. There are parallels with the McDonalds fast-food chain, although the supply of reliable and secure banking services is clearly more complicated than selling hamburgers. Apart from the trail-blazing approach of Citibank, it is worth wondering whether access to a global network of cash machines makes very much difference to customers and is really a selling point for the attraction of new account holders; the vast majority will only require access to banking services within their normal sphere of work and leisure activity. In view of these kinds of consideration and their potential impact on the costs and risks of global banking it is perhaps unlikely that many other banks will follow Citibank's example in the foreseeable future.

Unfulfilled Potential of Telecommunications?

While innovations in information technology will continue to influence the production, consumption and delivery of services, it remains the case that the potential of the technology already available has yet to be fully realized (Miles, 1991). Services embodied in human resources are an important and unavoidable method of production and delivery, and this often involves extensive and expensive travel by highly (as well as not so highly) paid personnel using the world's airlines. But there are alternatives. One such is video-conferencing, whereby a moving picture of the person being spoken to is transmitted simultaneously with the voice transmission. This has been available for some time and is now becoming more cost effective as the technology improves, usage increases and the cost of business travel by air rises. A two-hour video conference using specialist accommodation and perhaps involving three executives costs around £2000 (1991 prices); this is considerably less than the cost, both financial and temporal, of sending just one executive to a face-to-face meeting on the other side of the Atlantic or the Pacific. Service MNEs with large travel budgets could spend a fraction of the money on an in-house video conferencing facility and very quickly pay off the investment in reduced travel expenses. The scope for adopting this approach has been increased recently by agreement between the major Japanese, US and European manufacturers on an international telecommunications standard for these systems.

In Europe the replacement of traditional analogue phone circuits with integrated service digital networks (ISDNs) has allowed several telecommunications operators to embark on pan-European trials of videophones. Some 250 businesses in six countries were involved in 1991; although initial unit costs will be high (£5000 per stand-alone videophone) they will fall quite quickly as the service develops (Price, 1991). A computer board is being developed which users can slot into their desktop personal computers to receive pictures on the VDU screen of television quality while talking on the telephone.

Innovations in the applications of information and telecommunications technology are therefore set to continue. They are worth noting because their ultimate effect, threatened for many years but yet to be manifested, may be to replace a global service economy dependent on intensive transactional flows (people, information, electronic messages etc.) between a small number of large, highly centralized control points with a more dispersed telecommunications-dependent network of transactions. Flows of human capital will be far less significant although still necessary for the conduct of high-level negotiations and information exchange. Investment in, and access to, telecommunications infrastructure will be even more vital and this will still make it difficult for the less developed economies to participate fully and on an equitable basis in the global service economy.

The revolution in computer and telecommunications technology has also generated much speculation about the potential for teleworking or telecommuting. Digital telephone exchanges, modem links and powerful but affordable desktop and laptop computers enable individuals to work from home or small workstations near home rather than in the large office buildings in downtown or suburban office parks. Some of the geographical consequences of the globalization of services outlined in this volume would clearly be very different if large numbers of workers volunteered to adopt remote working or were required to do so by large corporate employers well aware of the cost savings to be gained from avoiding the high cost of providing office space in traditional high-density, competitive downtown locations. Many of the routine functions performed in the offices of a wide range of business, professional and transport services, for example, do not need to be undertaken in places where all the workers involved are grouped together at one location. Provided that they can receive and transmit work electronically and respond to requests and commands via telephone links there is no reason why they should be based at home. This is also true for less routine tasks connected with sales, marketing or service support, for example, where those involved either spend more time away from than actually working at their desks or are able to give telephone advice and assistance with, for example, computer software problems, without the need to be located in a particular place. Taken to its logical conclusion there would be only limited demand for the most contact-intensive service workers in strategic planning or financial management, for example, to be located in the urban agglomerations that have been the principal beneficiaries of the internationalization of services.

Although it is possible to point to numerous individual examples that illustrate the development of teleworking (see for example Burch, 1991) their cumulative impact on the established pattern of national and international service industry location has yet to become significant. Perhaps 0.5m full-time workers, mainly in the services, were engaged in telework in the UK in 1990, with a further 1.5m operating on a part-time basis (Burch, 1991). This falls well short of the large-scale growth forecast by futurologists during both the 1970s and the 1980s. The principal reason is not difficult to identify: teleworking requires a different expectation on the part of individuals about the role of work, especially the workplace, in their social and psychological as well as economic welfare. It will take a generation or more for those entering the labour force to have different expectations of how and where they will provide their labour; it cannot be very easy for those used to working in environments involving regular contact with many people to suddenly find themselves working on their own or in small groups at neighbourhood teleworking centres. Although work can be monitored by visits from supervisors, teleworking requires a good deal of self-discipline and realistic target setting if distractions are not going to have an adverse effect on performance and productivity. Advocates of teleworking suggest that it raises employee productivity but there is still no hard evidence to support this assertion. It would seem that while the potential of teleworking is considerable there is still much reluctance on the part of many organizations to adopt it as a major operational component of their organizations. Up to the year 2000 and beyond it seems, then, that the world cities will not be seriously threatened by the dispersal presaged by teleworking. Of much greater concern to them will be their vulnerability to fluctuations in the business cycle, especially as service industries have for the first time in the early 1990s begun to reflect phases in the business cycle in a way that, in the past, has been the sole domain of manufacturing industry.

References

Alessandrini, S., Secchi, C. and Saviolo, S. (1991) Competition among urban areas for the location of business services: a European perspective. Paper presented at the RESER International Conference on New Spatial Perspectives on Services, Lyon, 12–13 September.

Allen, J. (1988) Service industries: uneven development and uneven knowledge, *Area*, 20, 15–22.

Anderson, W. B. (1982) How to stem the rising protectionism, *Business Week*, 8 March, 16.

Bannon, M. J. and Tarbatt, J. (1990) *Skill Shortages in the Dublin Area: The Business Services Report*. Dublin: Service Industries Research Centre, mimeo.

Barcet, A. (1988) The development of tertiary services in the economy, labour market and employment, *The Service Industries Journal*, 8, 39–66.

Baro, E., Miralles, C., Soy, A. and Ursa, Y. (1989) *The Service Sector in the Metropolitan Region of Barcelona*. Barcelona: Centre d'Estudes i Planificacio.

Barras, R. (1983) A simple theoretical model of the office development cycle, *Environment and Planning A*, 15, 1381–94.

Barry, M. E. and Warfield, C. L. (1988) The globalization of retailing, *Textile Outlook International*, January, 62–76.

Bateman, M. (1985) *Office Development: A Geographical Analysis*. London: Croom Helm.

Batty, M. (1991) Urban and information networks: the evolution and planning of new computer and communications infrastructures, in Brotchie, J. F., Hale, P. and Newton, P. W. (eds), *The Spatial Impact of Technological Change*, Beckenham: Croom Helm, 139–57.

Baumol, W. (1967) Macroeconomics of unbalanced growth, *American Economic Review*, 57, 415–26.

Baumol, W. (1987) *Productivity Policy and the Service Sector*. Washington DC: Discussion Paper 1, Fishman-Davidson Center for the Study of the Service Sector.

Bell, D. (1973) *The Coming of Post-Industrial Society: A Venture in Social Forecasting*. New York: Basic Books.

Beniger, J. (1986) *The Control Revolution: Technological and Economic Origins of the Information Society*. Cambridge, MA: Harvard University Press.

Bertrand, O. and Noyelle, T. J. (1986) *Changing Technology, Skills and Skill Formation in French, German, Japanese, Swedish and US Financial Service Firms: Preliminary Findings*. Paris: OECD Centre for Educational Research and Innovation.

Bertrand, O. and Noyelle, T. J. (1988) *Corporate Strategy and Human Resources: Technological Change in Banks and Insurance Companies in Five OECD Countries*. Paris: OECD.

Beyers, W. B. (1989) *The Producer Services and Economic Development in the United States: The Last Decade*. Seattle: final report by US Department of Commerce, Economic Development Administration Technical Assistance and Research Division.

Bhagwatti, J. (1991) *The World Trading System at Risk*. New York: Harvester-Wheatsheaf.

Bhatt, U. V. (1989) On participating in the international capital market, in UNCTAD, *Services and Development Potential: The Indian Context*, New York: United Nations, 151–188.

Bina, C. and Yaghmaian, B. (1991) Post-war accumulation and trans-nationalization of capital, *Capital and Class*, 43 (spring), 107–130.

Black, T. J., Roark, P. and Schwartz, G. (1986) *The Changing Office Workplace*. Washington, DC: Urban Land Institute.

Blackaby, F. (1978) *De-industrialization*. London: Heinemann.

Blanden, M. (1986) Bigger role for foreign banks in the City, *The Banker*, Nov., 69–124.

Blanden, M. (1990) Space to spare. *The Banker*, 140, 769, March, 69–72.

Blois, K. J. (1983) Service marketing: assertion or asset?, *The Service Industries Journal*, 3, 113–20.

Booz-Allen and Hamilton (1990) *International Diversification Strategies for Telecommunications Service Companies*. London: Booz-Allen and Hamilton.

Borchert, J. R. (1978) Major control points in American economic geography, *Annals, Association of American Geographers*, 68, 214–32.

Borchert, J. R., Bourne, L. S. and Sinclair, R. (1986) *Urban Systems in Transition*. Amsterdam: Netherlands Geographical Studies 16, Koninklijk Nederlands Aardrijkskundig Genostschap.

Bouska, J. and Cerny, M. (1991) The position of the service sector in the Czechoslovak economy. Paper presented at the RESER International Conference on New Spatial Perspectives on Services, Lyon, September.

Boyd-Barrett, O. (1989) Multinational news agencies, in Enderwick, P. (ed.), *Multinational Service Firms*, London: Routledge, 107–31.

Bressand, A. (1989) Access to networks and services trade: the Uruguay Round and beyond, in UNCTAD, *Services in the World Economy*, New York: United Nations, 215–49.

Bressand, A. and Nicolaidis, K. (eds) (1989) *Strategic Trends in Services*. New York: Harper and Row.

Brotchie, J. F., Batty, M., Hall, P. and Newton, P. (eds) (1991) *Cities of the 21st Century: New Technologies and Spatial Systems*. New York: Halsted Press.

Brotchie, J. F., Hall, P. and Newton, P. W. (eds) (1987) *The Spatial Impact of Technological Change*. Beckenham: Croom Helm.

Brunn, S. D. and Leinbach, T. R. (eds) (1991) *Collapsing Space and Time: Geographic Aspects of Communication and Information*. New York: Harper Collins.

Buckley, P. J. and Casson, M. (1976) *The Future of the Multinational Enterprise*. London: Macmillan.

Buckley, P. J. and Casson, M. (1985) *The Economic Theory of the Multinational Enterprise*. London: Macmillan.

Burch, S. (1991) *Teleworking: A Strategic Guide for Management*. London: Kogan Page.

Camagni, R. P. (1990) Réseau de villes: éléments pour une théorisation et une taxonomie (Network of cities: towards a taxonomy and a theory). Paper presented at a conference 'Métropoles en déséquilibre?', Lyon, 22–23 November.

Castells, M. (1989) *The Informational City: Information Technology, Economic Restructuring, and the Urban-Regional Process*. Oxford: Blackwell.

Caves, R. E. (1982) *Multinational Enterprise and Economic Analysis*. Cambridge: Cambridge University Press.

Cecchini, P. (1988) *The European Challenge 1992: The Benefits of the Single Market*. London: Wildwood House.

Central Statistical Office (1991) *Monthly Review of External Trade Statistics*. London: Her Majesty's Stationery Office.

Christaller, W. (1933) *Die Zentralen Orte in Süddeutschland*. Jena: Fischer. Translated by C. W. Berkin as *Central Places in Southern Germany*, Englewood Cliffs, NJ: Prentice-Hall, 1966.

Christopher, M. (1984) The strategy of customer service, *The Service Industries Journal*, 4, 205–13.

Church, A. (1990) Transport and urban regeneration in London docklands: a victim of success or a failure to plan?, *Cities*, 7, 289–303.

Clairmonte, E. and Cavanagh, J. (1984) Transnational corporations and services: the final frontier, *Trade and Development*, 5, 215–73.

Clarke, W. M. (1986) *How the City Works*. London: Watstone.

Coakley, J. (1992) London as an international financial centre, in Budd, L. and Whimster, S. (eds), *Global Finance and Urban Living: A Study of Metropolitan Change*, London: Routledge, 52–72.

Coakley, J. and Harris, L. (1983) *The City of Capital: London's Role as a Financial Centre*. Oxford: Blackwell.

Cohen, R. (1979) The changing transactions economy and its spatial implications, *Ekistics*, 46, 7–15.

Cohen, S. S. and Zysman, J. (1987) *Manufacturing Matters: The Myth of the Post-Industrial Economy*. New York: Basic Books.

Commission of the European Communities (1988) *The 'Cost of Non-Europe' for Business Services: Executive Summary*. London: Peat Marwick Mitchell.

Cooper, I., Bain, A., Donaldson, J. and Price, L. (1987) *New Financial Instruments*. London: Chartered Institute of Bankers.

Coopers and Lybrand Deloitte (1990) *London as a Financial Centre in Europe in the 1990s*. London: Coopers and Lybrand Deloitte with British Telecom.

Corey, K. E. (1982) Transactional forces and the metropolis, *Ekistics*, 299, 416–23.

(CSI) (1989) *Company Operations in Developing Countries*. Washington, DC: Coalition of Service Industries.

Cowell, D. W. (1983) International marketing in services, *The Service Industries Journal*, 3, 308–28.

Cuadrado, J. R. and Del Rio, C. (1989) Structural change and the evolution of the service sector in the OECD, *The Service Industries Journal*, 9, 439–68.

Daly, M. T. (1991) Transitional economic bases: from the mass production society to the world of finance, in Daniels, P. W. (ed.), *Services and Metropolitan Development: International Perspectives*, London: Routledge, 26–43.

Daniels, P. W. (1987) Foreign banks and metropolitan development: a comparison of London and New York, *Tijdschrift voor Economische en Sociale Geografie*, 78, 269–87.

Daniels, P. W. (1990) *Change and Transition in Metropolitan Areas: The Role of Tertiary Industries (Final Report)*. Paris: report prepared for the World Association of the Major Metropolises on behalf of the Tertiary Industries Working Group.

Daniels, P. W. and Bobe, M. (1990) *Information Technology and the*

Renaissance of the City of London Office Building. Portsmouth: Working Papers, no. 3, Service Industries Research Centre.

Daniels, P. W. and Bobe, J. (1991) *High Rise and High Risks: Office Development on Canary Wharf.*. Portsmouth: Working Papers no. 7, Service Industries Research Centre.

Daniels, P. W., Hutton, T. A. and O'Connor, K. (1992) The planning response to urban service sector growth, *Growth and Change*, 22, 3–26.

Daniels, P. W., Thrift, N. J. and Leyshon, A. (1989) Internationalization of professional producer services: accountary conglomerates, in Enderwick, P., (ed.), *Multinational Service Firms*, London: Routledge, 79–106.

Danzin, A. (1983) The nature of new office technology, in Otway, H. J. and Peltu, M. (eds), *New Office Technology: Human and Organizational Aspects*, London: Pinter.

Davis, E. and Latter, A. R. (1989) London as an international financial centre, *Bank of England Quarterly Bulletin*, 29, 516–28.

DeAnne, Julius (1990) *Global Companies and Public Policy: The Growing Challenge of Foreign Direct Investment*. London: Pinter.

de Bandt, J. (ed.) (1985) *Les Services dans les sociétés industrielles avancées*. Paris: Economica.

Debenham Tewson Research (1990) *Offices: Gloom and Doom*. London: Debenham Tewson Research.

Debenham Tewson Research (1991) *Central London Office Market, January 1991*. London: Debenham Tewson Research.

de Mateo, F. (1988) El sector de servicios en Mexico: un diagnostico preliminar, *Commercio Exterior*, 38, January.

de Smidt, M. (1992) International investments and the European challenge, *Environment and Planning A*, 24, 83–94.

Diacon, S. (1990) Strategies for the Single European Market: the options for insurers, *The Service Industries Journal*, 10, 197–211.

Dicken, P. (1992) *Global Shift* (2nd edn). London: Paul Chapman.

Dixon, H. (1990) A clearer line to markets abroad, *The Financial Times*, 6 March.

Dixon, H. and Staple, G. (1991) Telegeography: new perspective on patterns of power, *Financial Times*, 7 October.

Drennan, M. P. (1985) *Modelling Metropolitan Economies for Forecasting and Policy Analysis*. New York: New York University Press.

Drennan, M. P. (1991) The decline and rise of the New York economy, in Mollenkopf, J. H. and Castells, M. (eds), *Dual City: Restructuring New York*, New York: Russel Sage Foundation, 25–41.

Drennan, M. P. (1992) Gateway cities: the metropolitan source of US producer service exports, *Urban Studies*, 29, 217–35.

Duffy, F. (1986) Exploding the myths of New York, *The Banker*, 136, 109.

Duffy, F. and Henney, A. (1989) *The Changing City*. London: Bullstrode Press.

Dunning, J. H. (1981) *International Production and the Multinational Enterprise*. London: Allen and Unwin.

Dunning, J. H. (1989) Multinational enterprises and the growth of services: some conceptual and theoretical issues, *The Service Industries Journal*, 9, 5–39.

Dunning, J. H. and Norman, G. (1983) The theory of the multinational enterprise: an application to multinational office location, *Environment and Planning A*, 15, 675–92.

Dunning, J. H. and Norman, G. (1987) The location choice of offices of international companies, *Environment and Planning A*, 19, 613–31.

Durie, J. (1990) US poised to lift bank restrictions, *The Times*, 17 April.

Dutton, W. H., Blumler, J. G. and Kraemer, K. L. (1987) *Wired Cities: Shaping the Future of Communications*. Boston, MA: Hall.

Edvardsson, B., Edvinsson, L. and Nyström, H. (1993) Internationalization in service companies, *The Service Industries Journal*, 13, 80–97.

Elfring, T. (1988) *Service Employment in Advanced Economies*. London: Gower.

Elfring, T. (1989a) The main features and underlying causes of the shift to services, *The Service Industries Journal*, 9, 337–56.

Elfring, T. (1989b) New evidence on the expansion of service employment in advanced economies, *Review of Income and Wealth*, series 35, 4, 409–40.

Elfring, T. (1989c) Reasons for the accelerated expansion of business and professional services. Paper presented at the Fifth Annual Seminar on the Service Economy, Geneva, 29–31 May.

Enderwick, P. (1989a) Some economies of service-sector multinational enterprises, in Enderwick, P. (ed.), *Multinational Service Firms*, London: Routledge, 3–34.

Enderwick, P. (ed.) (1989b) *Multinational Service Firms*. London: Routledge.

Euromoney (1986) International equities (supplement), November, 2.

Evans, W. (1990) The European banking market post-1992, *The Service Industries Journal*, 10, 188–96.

Everard, J. A. (1987) *The OECD Codes and Declarations and Trade in Services*. Victoria, BC: Working Papers on Trade in Services, Institute for Research on Public Policy.

Fainstein, S. S. (1990) Economics, politics and development policy: the convergence of New York and London, *International Journal of Urban and Regional Research*, 14, 553–75.

Faust, P. (1989) Shipping services, in UNCTAD, *Trade in Services: Sectoral Issues*, New York: United Nations, 113–152.

Feagin, J. R. and Smith, M. P. (1987) Cities and the new international division of labour: an overview, in Smith, M. P. and Fagin, J. R. (eds), *The Capitalist City*, Oxford: Blackwell.

Feketekuty, G. (1988) *International Trade in Services: An Overview and Blueprint for Negotiations*. Cambridge, MA: Ballinger.

Financial Times (1987) International capital markets, *Financial Times*, 21 April.

Financial Times, (1991a) How costs of living compare around the globe, *The Financial Times*, 16 January.

Financial Times(1991b) Seoul to allow more access to services, *Financial Times*, 17 January.

Financial Times (1991c) World market share shrinks, *Financial Times*, 13 March.

Fisher, A. G. (1935) *The Clash of Progress and Serenity*. London.

Ford, L. R. (1992) Reading the skylines of American cities, *Geographical Review*, 82, 180–200.

Fourastie, J. (1949) *Le Grand Espoir du XXe siècle*. Paris.

Friedmann, J. (1986) The world city hypothesis, *Development and Change*, 17, 69–83.

Friedmann, J. and Wolff, G. (1982) World city formation: an agenda for research and action, *International Journal of Urban and Regional Research*, 6, 309–44.

Frost and Sullivan Inc. (1985) *Value-Added Networks in Europe*. London: Frost and Sullivan.

Fuchs, V. R. (1968) *The Service Economy*. New York: National Bureau of Economic Research.

Gadrey, J. (1988) Rethinking output in services, *The Service Industries Journal*, 8, 67–76.

GATT (1989) *International Trade 1988–89*. Geneva: General Agreement on Tariffs and Trade.

Geddes, P. (1915) *Cities in Evolution*. London: Association for Planning and Regional Reconstruction.

Gershuny, J. and Miles, I. (1983) *The New Service Economy: The Transformation of Employment in Industrial Societies*. London: Frances Pinter.

Gertler, M. (1988) The limits to flexibility: comments on the post-Fordist vision of production and its geography, *Transactions, Institute of British Geographers*, NS13, 414–432.

Gibb, J. M. (1982) Information transfer in Europe, in Hills, P. J. (ed.), *Trends in Information Transfer*. London, Pinter, 47–63.

Gibbs, M. and Hayashi, M. (1989) Sectoral issues and the multinational framework for trade in services: an overview, in UNCTAD, *Trade in Services: Sectoral Issues*, New York: United Nations, 1–48.

Goldberg, M. A., Helsley, R. W. and Levi, M. D. (1987) *The Evolution of International Financial Centres: Comparative Analysis, Agglomeration Economies and Policy Implications*. Vancouver: Faculty of Commerce and Business Administration, University of British Columbia, mimeo.

Goldberg, M. A., Helsley, R. W. and Levi, M. D. (1991) The growth of international financial services and the evolution of international financial centres: a regional and urban economic approach, in Daniels, P. W. (ed.), *Services and Metropolitan Development: International Perspectives*, London: Routledge, 44–65.

Gottman, J. (1970) Urban centrality and the interweaving of quarternary activities, *Ekistics*, 29, 322–31.

Gottman, J. (1983) *The Coming of the Transactional Metropolis*. College Park: Institute for Urban Studies, University of Maryland.

Greenfield, H. I. (1966) *Manpower and the Growth of Producer Services*. New York: Columbia University Press.

Grubel, H. G. (1987) All traded services are embodied in goods or people, *The World Economy*, 10, 319–30.

Grubel, H. G. and Walker, M. A. (1989) *Service Industry Growth: Causes and Effects*. Vancouver: Fraser Institute.

Guile, P. and Quinn, J. B. (eds) (1988) *Technology in Services: Policies for Growth, Trade and Employment*. Washington DC: National Academy Press.

Hall, P. G. (1966) *The World Cities*. London: Weidenfield and Nicholson.

Hall, P. G. (1991a) Moving information: a tale of four technologies, in Brotchie, J. F., Batty, M., Hall, P. and Newton, P. (eds), *Cities of the 21st Century*, New York: Halsted Press, 1–21.

Hall, P. G. (1991b) The restructuring of urban economies: integrating urban and sectoral policies, in Fox-Przeworski, J., Goddard, J. B. and de Jong, M. *Urban Regeneration in a Changing Economy*. Oxford: Clarendon Press, 7–23.

Hamilton, J. D. (1986) *Stockbroking Tomorrow*. London: Macmillan.

Harrington, J. W. (1989) *Trade in Services between the U.S. and Canada: Status and Prospects*. SUNY, Buffalo: Canada–United States Trade Centre, Occasional Paper no. 2.

Harvey, D. (1982) *The Limits to Capital*. Oxford: Blackwell.

Hepworth, M. E. (1989) *Geography of the Information Economy*. London: Belhaven.

Hill, M. R. (1989) Soviet and Eastern European service multinationals, in

Enderwick, P. (ed.), *Multinational Service Firms*, London: Routledge, 155–99.

Hindley, B. and Smith, A. (1984) Comparative advantage and trade in services, *The World Economy*, 7, 377–81.

Hirsch, S. (1986) International transactions in services and in service intensive goods (mimeo).

Hirschorn, L. (1988) The post-industrial economy: labour skills and the new mode of production, *The Service Industries Journal*, 8, 19–38.

House of Commons Trade and Industry Committee (1988) *Information Technology* (2 vols). London: HMSO.

Howells, J. (1988) *Economic, Technological and Locational Trends in European Services*. Aldershot: Avebury.

Howland, M. (1992) Technological change and spatial restructuring of data entry and processing services, *Technological Change and Social Formation* (forthcoming).

Hugill, P. (1990) The international investor monies, in *Comment 90*, London: Knight Frank Rutley, 24–5.

Illeris, S. (1989) *Services and Regions in Europe*. Aldershot: Avebury.

International Institute of Communications (1990) *The Global Telecommunications Traffic Boom*. London: International Institute of Communications.

International Institute of Communications (1991) *The Global Telecommunications Traffic Report, 1991*. London: International Institute of Communications.

International Labour Office (1987) *Year Book of International Labour Statistics*. Geneva: International Labour Office.

Janelle, D. G. (1991) Global interdependence and its consequences, in Brunn, S. D. and Leinbach, T. R. (eds), *Collapsing Space and Time*, New York: Harper Collins, 49–81.

Jones, G. (ed.) (1992) *Multinational and International Banking*. Aldershot: Elgar.

Jones Lang Wootton (1990) *The City Office Review 1980–89: A Decade of Change*. London: Jones Lang Wootton.

Jones Lang Wootton (1991) *The Decentralization of Offices from Central London: An Annual Survey*. London: Jones Lang Wootton.

Katouzian, M. A. (1970) The development of the service sector: a new approach, *Oxford Economic Papers*, 22.

Kellerman, A. (1985) The evolution of service economies: a geographical perspective, *The Professional Geographer*, 37, 133–43.

Kindleberger, C. P. (1958) *Economic Development*. New York: McGraw-Hill.

Kindleberger, C. P. (1974) *The Formation of Financial Centres: A Study in*

Comparative Economic History. Princeton, NJ: Princeton University Press.

Kirn, T. J. (1987) Growth and change in the service sector of the U.S.: a spatial perspective, *Annals, Association of American Geographers*, 77, 353–72.

Knight, R. V. (1989) City development in advanced industrial societies, in (eds), *Cities in the 21st Century*, New York: Praeger.

Knight, R. V. and Gappert, G. (1989) *Cities in a Global Society*. New York: Sage.

Knirsch, P. (1983) *East European Firms in the Federal Republic of Germany and Austria*. Geneva: Institute for Research and Information on Multinationals.

Knox, P. and Agnew, J. (1989) *The Geography of the World Economy*. London: Edward Arnold.

Kravis, I. B. and Lipsey, R. E. (1983) *Toward an Explanation of Natural Price Levels*. Princeton, NJ: Princeton Studies in International Finance, 52.

Krommenacker, R. J. (1984) *World-Traded Services: The Challenge for the Eighties*. Dedham, MA: Artech House.

Lambert, R. (1986) 24 hour markets: more products now need round the clock trading, *Financial Times*, 27 October.

Langdale, J. V. (1985) Electronic funds transfer and the internationalization of the banking and finance industry, *Geoforum*, 16, 1–13.

Langdale, J. V. (1989) The geography of international business telecommunications: the role of leased networks, *Annals, Association of American Geographers*, 79, 501–22.

Langdale, J. V. (1991) Telecommunications and international transactions in information services, in Brunn, S. D. and Leinbach, T. R. (eds), *Collapsing Space and Time*, New York: Harper Collins, 193–214.

Lascelles, D. (1991) Banking on boulevards to build the business, *Financial Times*, 24 July.

Lecraw, D. J. (1989) Third World multinationals in the service industries, in Enderwick, P. (ed.), *Multinational Service Firms*, London: Routledge, 200–12.

Levich, R. M. and Walter, I. (1989) The regulation of global financial markets, in Noyelle, T. (ed.), *New York's Financial Markets: The Challenge of Globalization*, Boulder: Westview Press, 51–90.

Levitan, S. A. (1985) Services and long term structural change, *Economic Impact*, 54, autumn.

Levitt, T. (1976) The industrialization of service, *Harvard Business Review*, 54, 63–74.

Leyshon, A., Daniels, P. W. and Thrift, N. J. (1987) *Internationalization of*

Professional Producer Services: the Case of Large Accountancy Firms. Portsmouth: Working Papers on Producer Services no. 3, University of Bristol and Service Industries Research Centre, Portsmouth Polytechnic.

Leyshon, A. and Thrift, N. J. (1992) Liberalisation and consolidation: the Single European Market and the remaking of European financial capital, *Environment and Planning A*, 24, 49–81.

Leyshon, A., Thrift, N. J. and Daniels, P. W. (1987a) *The Urban and Regional Consequences of the Restructuring of World Financial Markets: The Case of the City of London.* Portsmouth: Working Papers on Producer Services no. 4, University of Bristol and Service Industries Research Centre, Portsmouth Polytechnic.

Leyshon, A., Thrift, N. J. and Daniels, P. W. (1987b) *Large Commercial Property Firms in the UK: The Operational Development and Spatial Development of General Practice Firms of Chartered Surveyors.* Portsmouth: Working Papers on Producer Services no. 5, University of Bristol and Service Industries Research Centre, Portsmouth Polytechnic.

London Planning Advisory Committee (1991) *London: World City.* London: Her Majesty's Stationery Office.

Maciejewicz, J. and Monkiewicz, J. (1989) Changing role of services in the socialist countries of eastern Europe, *The Service Industries Journal*, 9, 384–98.

Mansfield, E. and Romeo, A. (1980) Technology transfer to overseas subsidiaries by US based firms, *Quarterly Journal of Economics*, 94, 737–50.

Markusen, J. E. (1989) Service trade by the multinational enterprise, in Enderwick, P. (ed.) *Multinational Service Firms*, London: Routledge, 35–59.

Marshall, J. N., Damesick, P. and Wood, P. (1987) Understanding the location and role of producer services in the UK, *Environment and Planning A*, 19, 575–95.

Marshall, J. N., Wood, P., Daniels, P. W., McKinnon, A., Bachtler, J., Damesick, P., Thrift, N., Gillespie, A., Green, A. and Leyshon, A. (1988) *Services and Uneven Development.* Oxford: Oxford University Press.

McKinsey and Company (1987) *Systems Technology and the United States Commercial Banking Industry.* New York: McKinsey and Company.

McKinsey Global Institute (1992) *Service Sector Productivity.* Washington, DC: McKinsey and Company.

McMillan, C. H. (1979) Growth of external investments by OECD countries, *The World Economy*, 2, 363–86.

McUsic, M. (1987) US manufacturing: any cause for alarm?, *New England Economic Review*, January/February.

Mendelsohn, M. S. (1980) *Money on the Move: The Modern International Capital Market*. New York: McGraw-Hill.

Miles, I. (1991) Telecommunications: abolishing space and reinforcing distance, in Brotchie, J. F., Batty, M., Hall, P. and Newton, P. (eds), *Cities of the 21st Century*, New York: Halsted Press, 73–93.

Millward, R. (1988) The UK services sector, productivity change and the recession in long-term perspective, *The Service Industries Journal*, 8, 263–76.

Modivel, S.K. (1989) Negotiating a multilateral framework in services: Montreal and after, in UNCTAD, *Services and Development Potential: The Indian Context*, New York: United Nations, 113–49.

Mody, B. (1989) *Export Promotion, Telecommunication and Regional Development in the Periphery: the Case of the Computer Software Sector in India*. East Lansing: Urban Affairs Programs and the Department of Telecommunications, Michigan State University, mimeo.

Momigliano, F. and Siniscalco, D. (1982) The growth of service employment: a reappraisal, *Banca Nacionale de Lavoro Quarterly Review*, September, 269–306.

Moran, M. (1991) *The Politics of the Financial Services Revolution*. London: Macmillan.

Moss, M. L. (1986) Telecommunications systems and large world cities: a case study of New York, in Lipman, A. D., Sugarman, A. D. and Cushman, R. F. (eds), *Teleports and the Intelligent City*, New York: Dow Jones, Irvin, 379–97.

Moss, M. L. (1987) Telecommunications, world cities and urban policy, *Urban Studies*, 24, 534–46.

Moss, M. L. (1988) Telecommunications: shaping the future, in Sternlieb, G. and Hughes, J. W. (eds), *America's New Market Geography: Nation, Region and Metropolis*, New Brunswick, NJ: Centre for Urban Policy Research, Rutgers.

Moss, M. (1989) The information city in the global economy. Paper presented at the Third International Workshop on Innovation, Technological Change and Spatial Impacts, Cambridge, 3–5 September.

Moss, M. L. (1991) The information city in the global economy, in Brotchie, J. F., Batty, M., Hall, P. and Newton, P. (eds), *Cities of the 21st Century*, New York: Halsted Press, 181–90.

Moss, M. L. and Brion, J. (1988) *Back Offices and Data Processing Facilities in New York State*. New York: New York University Urban Research Centre.

Moss, M. L. and Dunau, A. (1986) *The Location of Back Offices: Emerging*

Trends and Development Patterns. New York: New York University Urban Research Centre.

Nesporova, A. (1990) *The New Role of Services in the Czechoslovak Economy.* Prague: Institute for Forecasting.

Netzer, D. (1992) The economy of the New York Metropolitan Region, then and now, *Urban Studies,* 29, 251–8.

Newton, P. J. (1989) Telematic underpinnings of the information economy in Brotchie, J. F., Batty, M., Hall, P. and Newton, P. (eds), *Cities of the 21st Century,* New York: Halsted Press, 95–126.

Noyelle, T. J. (1985) *New Technologies and Services: Impacts on Cities and Jobs.* College Park: The University of Maryland Institute for Urban Studies.

Noyelle, T. J. (1989) New York's competitiveness, in Noyelle, T. J. (ed.), *New York's Financial Markets: The Challenge of Globalization,* Boulder: Westview Press, 51–90.

Noyelle, T. J. and Peace, P. (1991) Information industries: New York's new export base, in Daniels, P. W. (ed.), *Services and Metropolitan Development: International Perspectives,* London: Routledge, 284–304.

Noyelle, T. J. and Stanback, T. M. Jr (1984) *The Economic Transformation of American Cities.* Totowa, NJ: Rowman and Allanheld.

Noyelle, T. J. and Stanback, T. M. (1988) *The Postwar Growth of Services in Developed Economies.* Geneva: report to United Nations Commission on Trade and Development.

Nusbaumer, J. (1987) *Services in the Global Market.* Boston: Kluwer.

Ochel, W. and Wegner, M. (1987) *Service Economies in Europe. Opportunities for Growth.* London: Pinter.

OECD (1981) *Recent International Direct Investment Trends.* Paris: OECD.

OECD (1986) *Financial Market Trends.* Paris: OECD.

OECD (1990) *Trade in Services and Developing Countries.* Paris: OECD.

O'Farrell, P. N. and Hitchens, D. M. (1990) Producer services and regional development: key conceptual issues of taxonomy and quality measurement, *Regional Studies* 24, 163–71.

Peach, W. (1985) Changing patterns in the City, *Estates Gazette,* 19 January, 33–4.

Perna, N. S. (1987) The shift from manufacturing to services, *New England Economic Review,* January/February.

Perry, M. (1990) The internationalization of advertising, *Geoforum,* 22, 35–50.

Petit, P. (1987) Internationalization of services and new forms of competition. Paper presented at the Sixth International Conference of Europeanists, Washington DC, 30 October to 1 November.

Piore, M. J. and Sabel, C. F. (1984) *The Second Industrial Divide: Possibilities for Prosperity*. New York: Basic Books.

Pipe, G. R. (1989) Telecommunications services: considerations for developing countries in Uruguay Round negotiations, in UNCTAD, *Trade in Services: Sectoral Issues*, New York: United Nations, 49–107.

Plender, J. and Wallace, P. (1985) *The Square Mile*. London: Century.

Porat, M. (1977) *The Information Economy: Definition and Measurement*. Washington DC: US Department of Commerce, Office of Telecommunications Special Publication 77–12(i).

Porter, M. E. (1986) *Competition in Global Industries*. Boston, MA: Harvard Business School Press.

Pred, A. R. (1973) The growth and development of systems of cities in advanced economies, in Pred, A. R. and Tornquist, G. (eds), *Systems of Cities and Information Flows*, Lund: Gleerup, 9–82.

Pred, A. R. (1977) *City-Systems in Advanced Economies*. London: Hutchinson.

Price, C. (1991) Talking pictures, *Financial Times*, 13 September.

Price, K. A. (1986) *The Global Financial Village*. London: Banking World.

Ramsthorn, A. (1989) The top ten international design networks, *Financial Times*, 22 February.

Reed, H. C. (1981) *The Pre-eminence of International Financial Centres*. New York: Praeger.

Richardson, J. B. (1987) Approaches to services in trade theory, in Giarini, O. (ed.), *The Emerging Service Economy*, Oxford; Pergamon.

Richardson, J. B. (1988) Towards an agreement on trade in services. Paper prepared for 'Services Trade and Development: Towards a Framework', Centre for Applied Studies in International Negotiations, Geneva, May.

Richmond, D. E. (1989) *Report on the Growth and Importance of Tertiary Industry in Metropolitan Toronto*. Toronto: Municipality of Metropolitan Toronto, mimeo.

Riddle, D. I. (1986) *Service-Led Growth: The Role of the Service Sector in World Development*. New York: Praeger.

Riddle, D. I. (1989) The role of services in economic development: problems of definition and measurement, in UNCTAD, *Services and Development Potential: The Indian Context*, New York: United Nation, 33–45.

Rimmer, P. J. (1991) The global intelligence corporations and world cities: engineering consultancies on the move, in Daniels, P. W. (ed.), *Services and Metropolitan Development: International Perspectives*, London: Routledge, 66–106.

Ross, R. and Trachter, K. (1983) Global cities and global classes: the peripheralization of labour in New York city, *Review*, 6, 393–41.

Rugman, A. M. (1979) *International Diversification and the Multinational Enterprise*. Lexington, MA: Lexington Books.

Sapir, A. and Lutz, E. (1981) *Trade in Services: Economic Determinants and Development-Related Issues*. Washington DC: World Bank Staff Working Paper no. 480, World Bank.

Sassen, S. (1992) *The Global City*. Princeton, NJ: Princeton University Press.

Sauvant, K. P. and Zimney, Z. (1989) Foreign direct investment in services: the neglected dimension in international service negotiations, in UNCTAD, *Services and Development Potential: The Indian Context*, New York: United Nations, 71–112.

Scanlon, R. (1990) *The Role of the Service Industries in the Economy of New York City and its Metropolitan Region, 1977–87*. New York: Port Authority of NY–NJ, mimeo.

Schneider, O. (1991) The problems of the development of the service sector in Czechoslovakia. Paper presented at the RESER International Conference on New Spatial Perspectives on Services, Lyon, 12–13 September.

Schoenberger, E. (1988) From Fordism to flexible accummulation: technology, competition strategies and international location, *Environment and Planning D*, 6, 245–262.

Schwamm, H. and Merciai, P. (1985) *The Multinationals and the Services*. Geneva: IRM Multinational Report.

Scott, A. and O'Connor, K. (1991) The pattern of airline networks and airports in the Pacific Rim 1947–1990. Paper prepared for the Third Meeting of the Pacific Rim Council on Urban Development, Vancouver, October, mimeo.

Segebarth, K. (1990) Some aspects of international trade in services: an empirical approach, *The Service Industries Journal*, 10, 266–83.

Shelp, R. K. (1981) *Beyond Industrialization: The Ascendancy of the Global Service Economy*. New York: Praeger.

Singelmann, J. (1978) *From Agriculture to Services: The Transformation of Industrial Employment*. Beverley Hills, CA: Sage.

Smith, A. D. (1972) *The Measurement and Interpretation of Service Output Changes*. London: National Economic Development Office.

Srivastava, S. (1989) Computer software and data processing: export potential, in UNCTAD, *Services and Development Potential: The Indian Context*, New York: United Nations, 189–94.

Stanback, T. M. (1979) *Understanding the Service Economy*. Baltimore, MD: Johns Hopkins University Press.

Stanback, T. M. (1991) *The New Suburbanization: Challenge to the Central City*. Boulder: Westview Press.

Stanback, T. M., Bearse, P., Noyelle, T. and Karesek, R. (1981) *Services: The New Economy*. Totowa, NJ: Allanheld and Osman.

Stanback, T. M. and Noyelle, T. J. (1982) *Cities in Transition: Changing Job Structures in Atlanta, Denver, Buffalo, Phoenix, Columbus (Ohio), Nashville, Charlotte*. Tolowa, NJ: Allenheld and Osman.

Stern, R. M., Trezise, P. H. and Whalley, J. (eds) (1987) *Perspectives on a US–Canadian Free Trade Agreement*. Ottawa: Institute for Research on Public Policy.

The Times (1991a) A tale of two tax haven cities, *The Times*, 3 January.

The Times (1991b) Manufactured concern, *The Times*, 7 March.

Thrift, N. J. (1984) The internationalization of producer services and the genesis of a world city property market. Paper presented at Symposium on Regional Development Processes/Policies and the Changing International Division of Labour, Vienna, August, mimeo.

Thrift, N. J. (1986) *The 'Fixers': The Urban Geography of International Financial Capital*. Lampeter, University of Wales Department of Geography, mimeo.

Thrift, N. J. and Leyshon, A. (1992) In the wake of money: the City of London and the accumulation of value, in Budd, L. and Whimster, S. (eds), *Global Finance and Urban Living: A Study of Metropolitan Change*, London: Routledge, 282–311.

Thrift, N. J., Leyshon, A. and Daniels, P. W. (1987) *'Sexy Greedy': The New International Financial System, The City of London and the South East of England*. Portsmouth: Working Papers on Producer Services no. 8, University of Bristol and Service Industries Research Centre, Portsmouth Polytechnic.

Timberlake, M. (ed.) (1985) *Urbanization in the World Economy*. London: Academic Press.

Treadgold, A. D. and Davies, R. L. (1988) *The Internationalization of Retailing*. Harlow: Longman.

UNCTAD (1989a) *Services in the World Economy*. New York: United Nations (Trade and Development Report, 1988).

UNCTAD (1989b) *Services and Development Potential: The Indian Context*. New York: United Nations.

UNCTC (1988) *Transnational Corporations in World Development: Trends and Prospects*. New York: United Nations.

UNCTC (1989) *Foreign Direct Investment and Transnational Corporations in Services*. New York: United Nations.

UNCTC (1990) *Directory of the World's Largest Service Companies*. New York: Moody's Investor Services and UNCTC.

UNCTC (1991) *World Investment Report 1991: The Triad in Foreign Direct Investment*. New York: United Nations.

US Department of Commerce (1976) *Industrial Outlook Report*. Springfield, VA: National Technical Information Service.

US Department of Commerce (1985) *US Direct Investment Abroad: 1982 Benchmark Survey Data*. Washington DC: Government Printing Office.

Vandermerwe, S. and Chadwick, M. (1989) The internationalization of services, *The Service Industries Journal*, 9, 79–93.

Walker, R. (1985) Is there a service economy? The changing capitalist division of labour, *Science and Society*, 39, 42–83.

Warf, B. (1987) Service sector growth since the New York renaissance. Paper presented at the Annual Meeting of the Association of American Geographers, Portland, Oregon, 23–27 April.

Warf, B. (1991) The internationalization of New York services, in Daniels, P. W. (ed.), *Services and Metropolitan Development: International Perspectives*, London: Routledge, 245–64.

Williams, S. (1992) The coming of the groundscrapers, in Budd, L. and Whimster, S. (eds), *Global Finance and Urban Living*, London: Routledge, 246–59.

Williamson, O. E. (1979) Transaction cost economics: the governance of contractural relations, *The Journal of Law and Economics*, 22, 233–61.

Yannopoulos, G. N. (1983) The growth of transnational banking, in Casson, M. C. (ed.), *The Growth of International Business*, London: Allen and Unwin.

Zagor, K. (1991) Manhattan braces for thousands of job losses, *Financial Times*, 14 August.

Zukin, S. (1992) The city as a landscape of power: London and New York as global financial capitals, in Budd, L. and Whimster, S. (eds), *Global Finance and Urban Living*, London: Routledge, 195–233.

Further Reading

The terms of reference for the *Studies in Geography* series have not always made it possible to develop the individual themes in this book in the breadth and depth required by some readers. If this volume has acted as a primer to stimulate your interest in following up some of the ideas outlined in these chapters, the further reading identified below may be useful.

Allen, J. (1988) Service industries: uneven development and uneven knowledge, *Area*, 20, 15–22.

Bateman, M. (1985) *Office Development: A Geographical Analysis*. London: Croom Helm.

Bressand, A. and Nicolaidis, K. (eds) (1989) *Strategic Trends in Services*. New York: Harper and Row.

Brotchie, J. F., Batty, M., Hall, P. and Newton, P. (1991) *Cities of the 21st Century: New Technologies and Spatial Systems*. New York: Halsted Press.

Brunn, S. D. and Leinbach, T. R. (eds) (1991) *Collapsing Space and Time: Geographic Aspects of Communication and Information*. New York: Harper Collins.

Budd, L. and Whimster, S. (eds) (1992) *Global Finance and Urban Living: A Study of Metropolitan Change*. London: Routledge.

Castells, M. (1989) *The Informational City: Information Technology, Economic Restructuring, and the Urban-Regional Process*. Oxford: Blackwell.

Clairmonte, E. and Cavanagh, J. (1984) Transnational corporations and services: the final frontier, *Trade and Development*, 5, 215–73.

Daniels, P. W. (1985) *Service Industries: A Geographical Appraisal*. London: Methuen.

Daniels, P. W. (ed.) (1991) *Services and the Metropolitan Development: International Perspectives*. London: Routledge.

Dicken, P. (1992) *Global Shift*, 2nd edn. London: Paul Chapman Publishing.

Dunning, J. H. and Norman, G. (1987) The location choice of offices of international companies, *Environment and Planning A*, 19, 613–31.

Elfring, T. (1989a) The main features and underlying causes of the shift to services, *The Service Industries Journal*, 9, 337–56.

Elfring, T. (1989b) New evidence on the expansion of service employment in advanced economies, *Review of Income and Wealth*, Series 35, 4, 409–40.

Enderwick, P. (ed.) (1989b) *Multinational Service Firms*. London: Routledge.

Feketekuty, G. (1988) *International Trade in Services: An Overview and Blueprint for Negotiations*. Cambridge, Mass.: Ballinger.

Friedmann, J. (1986) The world city hypothesis, *Development and Change*, 17, 69–83.

Grubel, H. G. and Walker, M. A. (1989) *Service Industry Growth: Causes and Effects*. Vancouver: The Fraser Institute.

Guile, P. and Quinn, J. B. (eds) (1988) *Technology in Services: Policies for Growth, Trade and Employment*. Washington DC: National Academy Press.

Hepworth, M. E. (1989) *Geography of the Information Economy*. London: Belhaven.

Howells, J. (1988) *Economic, Technological and Locational Trends in European Services*. Aldershot: Avebury.

Knox, P. and Agnew, J. (1989) *The Geography of the World Economy*. London: Edward Arnold.

Krommenacker, R. J. (1984) *World-Traded Services: The Challenge for the Eighties*. Dedham, Mass.: Artech House.

Langdale, J. V. (1984) The geography of international business telecommunications: the role of leased networks, *Annals, Association of American Geographers*, 79, 501–22.

London Planning Advisory Committee (1991) *London: World City*. London: Her Majesty's Stationery Office.

Maciejewicz, J. and Monkiewicz, J. (1989) Changing role of services in the socialist countries of eastern Europe, *The Service Industries Journal*, 9, 384–98.

Moss, M. L. (1987) Telecommunications, world cities and urban policy, *Urban Studies*, 24, 534–46.

Noyelle, T. (ed) (1989) *New York's Financial Markets: The Challenge of Globalization*. Boulder: Westview Press.

Nusbaumer, J. (1987) *Services in the Global Market*. Boston: Kluwer Academic Publishers.

Ochel, W. and Wegner, M. (1987) *Service Economies in Europe. Opportunities for Growth*. London: Pinter.

OECD (1990) *Trade in Services and Developing Countries*. Paris: OECD.

Porter, M. E. (1986) *Competition in Global Industries*, Boston, Mass.: Harvard Business School Press.

Reed, H. C. (1981) *The Pre-eminence of International Financial Centres*. New York: Praeger.

Riddle, D. I. (1986) *Service-led Growth: The Role of the Service Sector in World Development*. New York: Praeger.

Sassen, S. (1992) *The Global City*. Princeton, NJ: Princeton University Press.

Schoenberger, E. (1988) From Fordism to flexible accumulation: technology, competition strategies and international location, *Environment and Planning D*, 6, 245–62.

Shelp, R. K. (1981) *Beyond Industrialization: The Ascendary of the Global Service Economy*. New York: Praeger.

UNCTAD (1989) *Services in the World Economy*. New York: United Nations (Trade and Development Report, 1988).

UNCTAD (1989) *Trade in Services: Sectoral Issues*, New York: United Nations.

UNCTC (1989) *Foreign Direct Investment and Transnational Corporations in Services*. New York: United Nations.

UNCTC (1990) *Directory of the World's Largest Service Companies*. New York: Moody's Investor Services and UNCTC.

Vandermerwe, S. and Chadwick, M. (1989) The internationalization of services, *The Service Industries Journal*, 9, 79–93.

Warf, B. (1989) Telecommunications and the globalization of financial services, *Professional Geographer*, 41, 257–71.

Index

Page numbers in italics indicate references to figures.

Index compiled by Ann Barham